MW00616319

STRUCTURAL ENGINEERING FORMULAS

COMPRESSION · TENSION · BENDING · TORSION · IMPACT
BEAMS · FRAMES · ARCHES · TRUSSES · PLATES
FOUNDATIONS · RETAINING WALLS · PIPES AND TUNNELS

ILYA MIKHELSON, PH.D.

ILLUSTRATIONS BY LIA MIKHELSON, M.S.

MCGRAW-HILL

NEW YORK CHICAGO SAN FRANCISCO LISBON LONDON MADRID
MEXICO CITY MILAN NEW DELHI SAN JUAN SEOUL
SINGAPORE SYDNEY TORONTO

The McGraw-Hill Companies

Library of Congress Cataloging-in-Publication Data

Mikhelson, Ilya.

Structural engineering formulas / Ilya Mikhelson.

p. cm.

ISBN 0-07-143911-0

1. Structural engineering—Mathematics. 2. Mathematics—Formulae. I. Title.

TA636.M55 2004

624.1'02'12—dc22 2004044803

2 3 4 5 6 7 8 9 0 DOC/DOC 0 1 0 9 8 7 6 5

ISBN 0-07-143911-0

The sponsoring editor for this book was Larry S. Hager, the editing supervisor was Stephen M. Smith, and the production supervisor was Sherri Souffrance. The art director for the cover was Margaret Webster-Shapiro.

Printed and bound by RR Donnelley.

This book is printed on recycled, acid-free paper containing a minimum of 50% recycled, de-inked fiber.

To my wife and son

CONTENTS

CONTENTS

CONTENTS

CONTENTS

PREFACE

This reference book is intended for those engaged in an occupation as important as it is interesting—design and analysis of engineering structures. Engineering problems are diverse, and so are the analyses they require. Some are performed with sophisticated computer programs; others call only for a thoughtful application of ready-to-use formulas. In any situation, the information in this compilation should be helpful. It will also aid engineering and architectural students and those studying for licensing examinations.

Ilya Mikhelson, Ph.D.

ACKNOWLEDGMENTS

Deep appreciation goes to Mikhail Bromblin for his unwavering help in preparing the book's illustrations for publication.

The author would also like to express his gratitude to colleagues Nick Ayoub, Tom Sweeney, and Davidas Neghandi for sharing their extensive engineering experience.

Special thanks is given to Larry Hager for his valuable editorial advice.

INTRODUCTION

Analysis of structures, regardless of its purpose or complexity, is generally performed in the following order:

- Loads, both permanent (dead loads) and temporary (live loads), acting upon the structure are computed.
- Forces (axis forces, bending moments, shears, torsion moments, etc.) resulting in the structure are determined.
- Stresses in the cross-sections of structure elements are found.
- Depending on the analysis method used, the obtained results are compared with allowable or ultimate forces and stresses allowed by norms.

The norms of structural design do not remain constant, but change with the evolving methods of analysis and increasing strength of materials. Furthermore, the norms for design of various structures, such as bridges and buildings, are different. Therefore, the analysis methods provided in this book are limited to determination of forces and stresses. Likewise, the included properties of materials and soils are approximations and may differ from those accepted in the norms.

All the formulas provided in the book for analysis of structures are based on the elastic theory.

About the Author

Ilya Mikhelson, Ph.D., has over 30 years' experience in design, research, and teaching design of bridges, tunnels, subway stations, and buildings. He is the author of numerous other publications, including: *Precast Concrete for Underground Construction*, *Tunnels and Subways*, and *Building Structures.*

1. STRESS and STRAIN

Methods of Analysis

N O T E S

Tables 1.1–1.12 provide formulas for determination of stresses in structural elements for various loading conditions. To evaluate the results, it is necessary to compare the computed stresses with existing norm requirements.

STRESS and STRAIN

TENSION and COMPRESSION

1.1

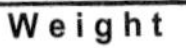

Weight

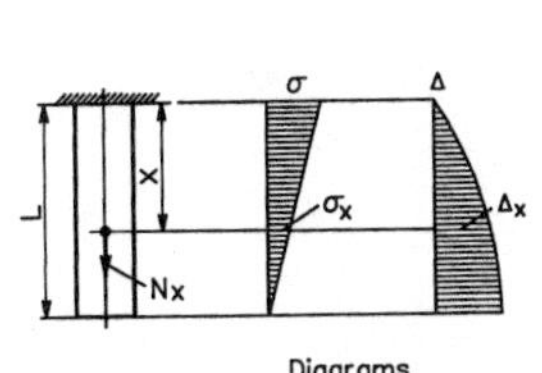

Axial force: $N_x = \gamma A(L-x)$,

γ = unit volume weight,

A = cross - sectional area.

Stresses: $\sigma_x = \dfrac{N_x}{A} = \gamma(L-x), \quad \sigma_{x=0} = \gamma L, \quad \sigma_{x=L} = 0.$

Deformation:

$$\Delta_x = \frac{\gamma x}{2E}(2L-x), \quad \Delta_{x=0} = 0, \quad \Delta_{x=L} = \frac{\gamma L^2}{2E} = \frac{W^2 L}{2EA}$$

$W = \gamma AL$ = weight of the beam

E = Modulus of elasticity

Axial force : tension, compression

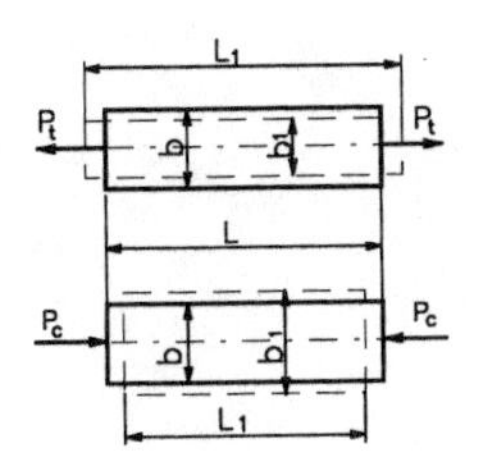

Stresses: $\sigma_t = \dfrac{P_t}{A}, \qquad \sigma_c = \dfrac{P_c}{A}.$

Deformation:

$$\Delta_L = L - L_1 \text{ (along)}, \quad \Delta_b = b - b_1 \text{ (cross)},$$

$$\varepsilon_L = \frac{\pm\Delta_L}{L}, \qquad \varepsilon_C = \frac{\mp\Delta_b}{b}.$$

$$\text{Poisson's ratio:} \quad \mu = \left[\frac{\varepsilon_c}{\varepsilon_L}\right].$$

Hooke's law $\sigma = E\varepsilon, \quad \varepsilon = \dfrac{\sigma}{E}$:

$$\Delta_L = \varepsilon_L L = \frac{\sigma}{E}L = \frac{P}{EA}L, \qquad \Delta_c = \varepsilon_c b = \frac{\mu\sigma}{E}b = \frac{\mu P}{EA}b.$$

Temperature

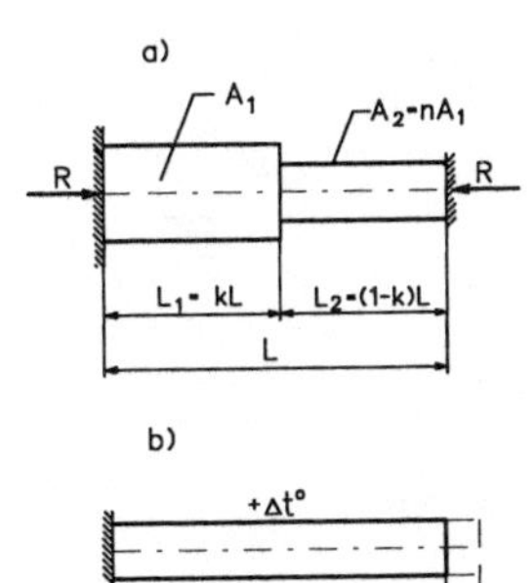

Case a/

$$\text{Reaction:} \quad R = \frac{\alpha \cdot \Delta t^0 EA}{k + \dfrac{1-k}{n}}, \qquad n = \frac{A_2}{A_1}, \qquad k = \frac{L_1}{L}.$$

Axial force $N = -R$ (compression),

$$\text{Stresses:}\ \sigma_1 = -\frac{R}{A_1} = -\frac{\alpha \cdot \Delta t^0 E}{k + \dfrac{1-k}{n}}, \quad \sigma_2 = -\frac{R}{nA_1} = -\frac{\alpha \cdot \Delta t^0 E}{k(n-1)+1}.$$

For $A_1 = A_2$: $\sigma = \sigma_1 = \sigma_2 = -\alpha \cdot \Delta t^0 E$, $\Delta t^0 = T_0^0 - T_c^0$

Where T_o^0 and T_c^0 are original and considered temperatures.

α = coefficient of linear expansion

$\Delta t^0 > 0$ tension stress, $\Delta t^0 < 0$ compression stress.

Case b/

Deformation: $\Delta_L^t = \alpha \cdot \Delta t^0 L.$

NOTES

Tables 1.2 and 1.3a

Example. Bending

Given. Shape W 14×30, $L = 6\text{m}$

Area $A = 8.85\text{in}^2 = 8.85\times2.54^2 = 57.097\text{cm}^2$

Depth $h = 13.84\text{in} = 13.84\times2.54 = 35.154\text{cm}$

Web thickness $d = 0.270\text{in} = 0.270\times2.54 = 0.686\text{cm}$

Flange width $b = 6.730\text{in} = 6.730\times2.54 = 17.094\text{cm}$

Flange thickness $t = 0.385\text{in} = 0.385\times2.54 = 0.978\text{cm}$

Moment of inertia $I_z = 291\text{in}^4 = 291\times2.54^4 = 12112.3\text{cm}^4$

Section modulus $S = 42.0\text{in}^3 = 42.0\times2.54^3 = 688.26\text{cm}^3$

Weight of the beam $\omega = 30\text{Lb/ft} = 30\times4.448/0.3048 = 437.8\ \text{N/m} = 0.4378\ \text{kN/m}$

Load $P = 80\ \text{kN}$

Allowable stress (assumed) $[\sigma] = 196.2\ \text{MPa}$, $[\tau] = 58.9\ \text{MPa}$

Required. Compute : σ_{max} and τ_{max}

Solution.

$$M = \frac{\omega L^2}{8} + \frac{PL}{4} = \frac{0.4378\times6^2}{8} + \frac{80\times6}{4} = 121.97\ \text{kN}\cdot\text{m}$$

$$V = \frac{\omega L}{2} + \frac{P}{2} = \frac{0.4378\times6}{2} + \frac{80}{2} = 41.31\ \text{kN}$$

$$\sigma_{max} = \frac{M}{S} = \frac{121.97\times100(\text{kN}\cdot\text{cm})}{688.26(\text{cm}^3)} = 17.72\ \text{kN/cm}^2 = 177215.0\ \text{kN/m}^2 = 177.215\ \text{MPa} < 196.2\ \text{MPa}$$

$$\tau_{max} = \frac{V}{I_z d}\left[bt\left(\frac{h}{2} - \frac{t}{2}\right) + \frac{d\left(\frac{h}{2} - t\right)^2}{2}\right] = 1.890\ \text{kN/cm}^2 = 18900\ \text{kN/m}^2 = 18.9\ \text{MPa} < 58.9\ \text{MPa}$$

STRESS and STRAIN

BENDING

1.2

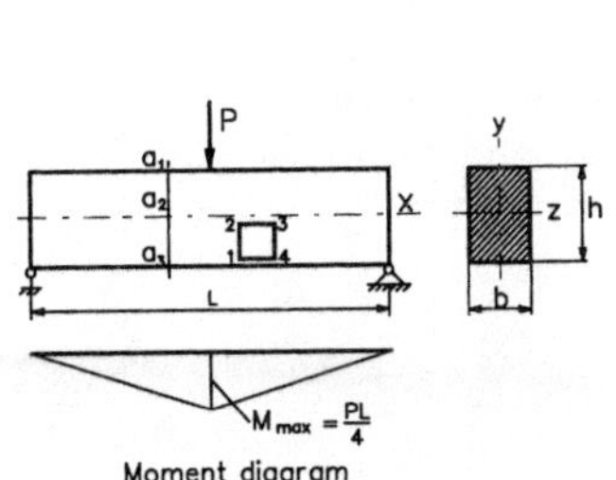

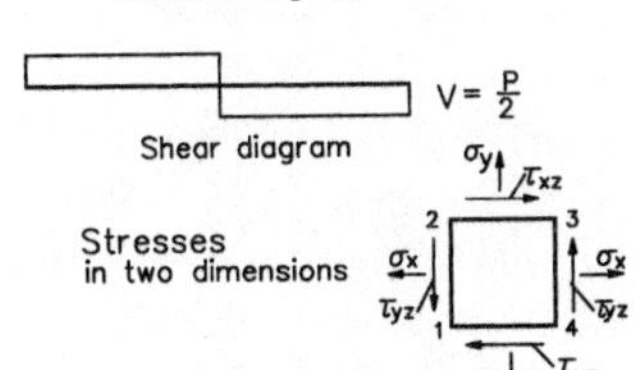

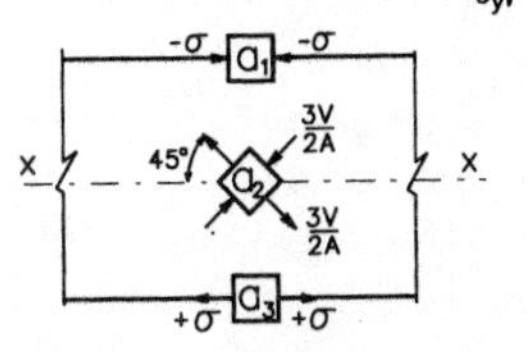

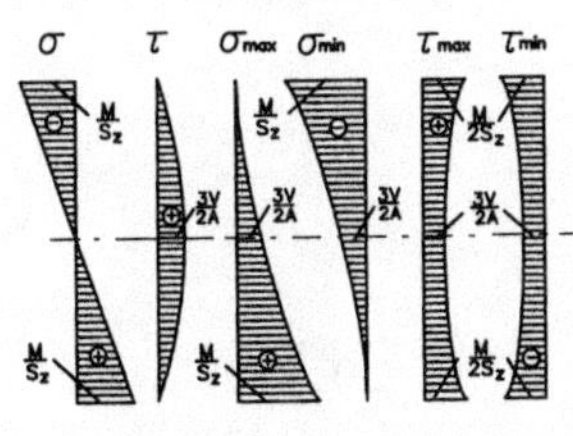

Stress diagrams

Bending stress: $\sigma = \frac{M}{I_z} \cdot y$

Shear stress: $\tau = \frac{VS}{I_z b}$

Stresses in x-y plane:

$$\sigma_y = 0, \quad \sigma_x = \sigma, \quad \tau_{xz} = \tau_{yz} = \tau$$

Principal stresses:

$$\sigma_{\substack{max \\ min}} = \frac{\sigma}{2} \pm \frac{1}{2}\sqrt{\sigma^2 + 4\tau^2}$$

Maximum shear (min) stresses:

$$\tau_{\substack{max \\ min}} = \pm \frac{1}{2}\sqrt{\sigma^2 + 4\tau^2}$$

The principal stresses and maximum (min) shear stresses lie at 45^0 to each other.

Stress diagrams
σ-diagram: $\sigma_{a_1} = +\frac{M}{S}, \quad \sigma_{a_2} = 0, \quad \sigma_{a_3} = -\frac{M}{S}.$
τ-diagram: $\tau_{a_1} = 0, \quad \tau_{a_2} = \frac{VS}{I_z b} = \frac{3V}{2A}, \quad \tau_{a_3} = 0.$
σ_{max}-diagram: $\sigma_{a_1} = +\frac{M}{S}, \quad \sigma_{a_2} = +\tau = +\frac{3V}{2A}, \quad \sigma_{a_3} = 0.$
σ_{min}-diagram: $\sigma_{a_1} = 0, \quad \sigma_{a_2} = -\tau = -\frac{3V}{2A}, \quad \sigma_{a_3} = -\frac{M}{S}.$
τ_{max}-diagram: $\tau_{a_1} = \tau_{a_3} = +\frac{\sigma}{2} = +\frac{M}{2S}, \quad \tau_{a_2} = +\tau = +\frac{3V}{2A}.$
τ_{min}-diagram: $\tau_{a_1} = \tau_{a_3} = -\frac{\sigma}{2} = -\frac{M}{2S}, \quad \tau_{a_2} = -\tau = -\frac{3V}{2A}.$

Note:
"+"- Tension
"−"- Compression

NOTES

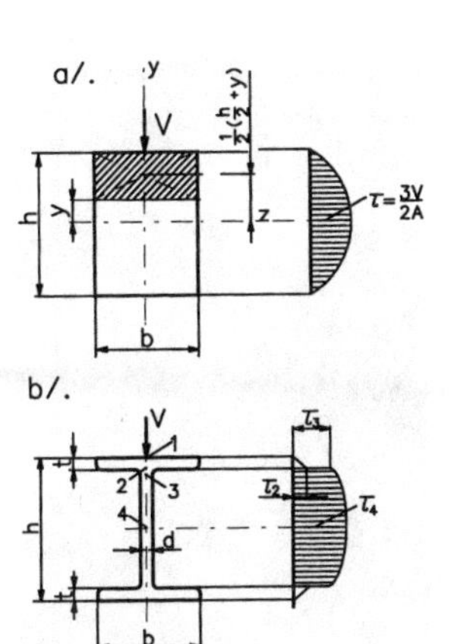

Shear stress: $\tau = \dfrac{VS}{I_z b}$

Case a/ $S_y = \dfrac{b}{2}\left(\dfrac{h}{2} - y\right)\left(\dfrac{h}{2} + y\right) = \dfrac{b}{2}\left(\dfrac{h^2}{4} - y^2\right),$

$$\tau = \frac{V \cdot \dfrac{b}{2}\left(\dfrac{h^2}{4} - y^2\right)}{\dfrac{bh^3}{12} \cdot b} = \frac{6V}{bh^3}\left(\frac{h^2}{4} - y^2\right)$$

for $y = \pm\dfrac{h}{2}$: $\tau = 0$, for $y = 0$: $\tau = \dfrac{3V}{2A}$

Case b/ $\tau_1 = 0,$

$$\tau_2 = \frac{V}{I_z b} bt\left(\frac{h}{2} - \frac{t}{2}\right), \quad \tau_3 = \frac{V}{I_z d} bt\left(\frac{h}{2} - \frac{t}{2}\right),$$

$$\tau_4 = \frac{V}{I_z d}\left[bt\left(\frac{h}{2} - \frac{t}{2}\right) + \frac{d\left(\dfrac{h}{2} - t\right)^2}{2}\right]$$

Bending in two directions

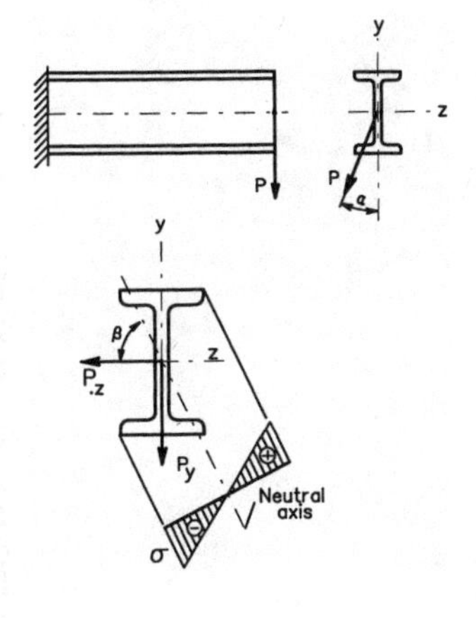

Bending moments.

Moment due to force P: $M = \sqrt{M_z^2 + M_y^2},$

$$M_z = M\cos\alpha, \quad M_y = M\sin\alpha,$$

$$\left[\frac{M_y}{M_z}\right] = [\tan\alpha]$$

For case shown: $M_z = P_y L\cos\alpha,\ M_y = P_z L\sin\alpha,$

$$M = PL$$

Stress:

$$\sigma = \pm M\left(\frac{y\cos\alpha}{I_z} + \frac{z\sin\alpha}{I_y}\right),$$

$$\sigma_{max} = \pm\frac{M}{S_z}\left(\cos\alpha + \frac{S_z}{S_y}\sin\alpha\right)$$

Neutral axis: $\tan\beta = \dfrac{I_z}{I_y}\tan\alpha.$

Deflection in direction of force P: $\Delta = \sqrt{\Delta_z^2 + \Delta_y^2},$

For case shown: $\Delta_z = \dfrac{P_z L^3}{3EI_y}, \quad \Delta_y = \dfrac{P_y L^3}{3EI_z}.$

NOTES

COMBINATION OF COMPRESSION (TENSION) and BENDING 1.4

Compression (Tension) and bending

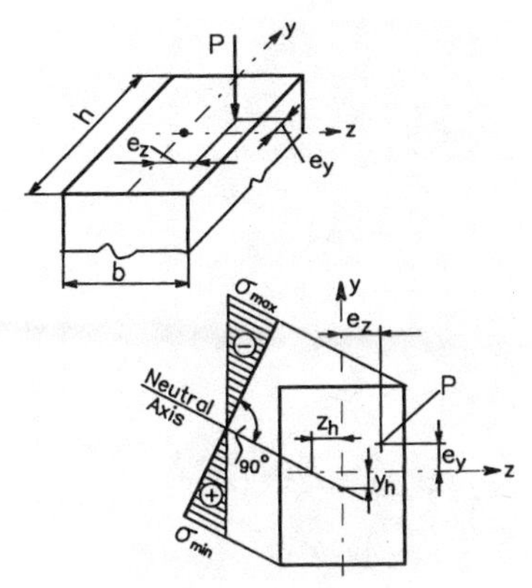

Stresses: $$\sigma = \frac{P}{A} \pm \frac{M_y}{I_y} z \pm \frac{M_z}{I_z} y,$$

$$\sigma_{\substack{max \\ min}} = \frac{P}{A} \pm \frac{M_y}{S_y} \pm \frac{M_z}{S_z}.$$

$$M_y = P \cdot e_z, \quad M_z = P \cdot e_y$$

$$I_y = \frac{h \cdot b^3}{12}, \quad I_z = \frac{b \cdot h^3}{12}, \quad S_y = \frac{h \cdot b^2}{6}, \quad S_z = \frac{b \cdot h^2}{6}$$

Neutral axis: $y_n = \frac{i_z^2}{e_y}, \quad z_n = \frac{i_y^2}{e_z}.$

$$i_z = \sqrt{I_z / A}, \quad i_y = \sqrt{I_y / A}, \quad A = b \cdot h$$

Buckling

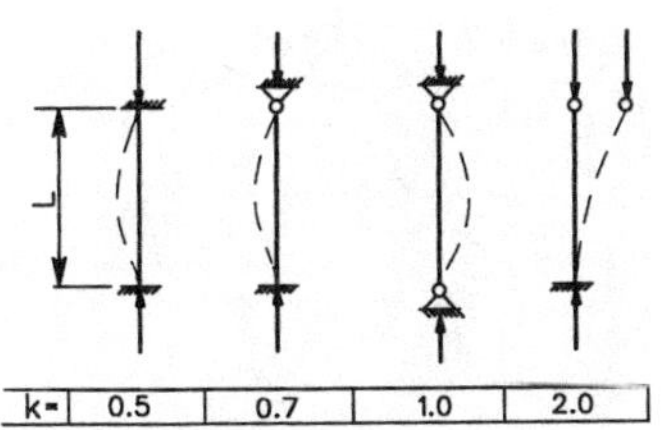

k=	0.5	0.7	1.0	2.0

Euler's formula:

$$P_e = \frac{\pi^2 EI}{(kL)^2} \quad \text{for} \quad \lambda_{min} \geq \pi \sqrt{\frac{E}{R_e}},$$

where R_e is the elastic buckling strength.

$$\lambda_{min} = \frac{kL}{i_{min}}, \quad \text{stress:} \quad \sigma_{max} \leq \frac{\pi^2 E}{\lambda_{min}^2}.$$

Axial compression (tension) and bending

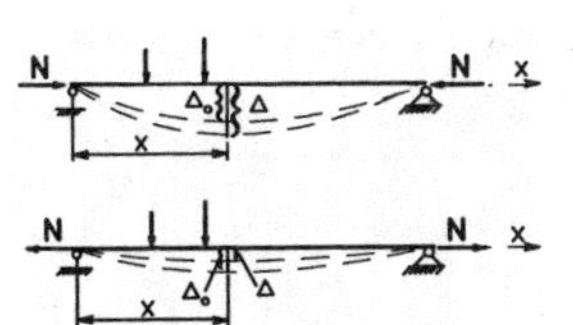

Stresses:

compression $$\sigma_{max} = \frac{N}{A} + \frac{M_0}{S_z} + \frac{N}{S_z} \cdot \frac{\Delta_0}{1 - \frac{N}{P_e}},$$

tension $$\sigma_{max} = \frac{N}{A} + \frac{M_0}{S_z} - \frac{N}{S_z} \cdot \frac{\Delta_0}{1 + \frac{N}{P_e}},$$

where: M_0 and Δ_0 = max. moment and max. deflection due to transverse loading.

NOTES

Table 1.5

Example. Torsion

Given. Cantilever beam, $L = 1.5\text{m}$, for profile see Table 1.5c

$h = 70\text{cm},\ h_1 = 30\text{cm},\ h_2 = 60\text{cm},\ h_3 = 40\text{cm},\ b_1 = 4.5\text{cm},\ b_2 = 2.5\text{cm},\ b_3 = 5.5\text{cm}$

Material: Steel, $G = 800\ \text{kN}/\text{cm}^2 = 8000\ (\text{MPa})$

Torsion moment $M_t = 40\ \text{kN}\cdot\text{m}$

Required. Compute τ_{max} and φ^0

Solution. $\frac{h_1}{b_1} = \frac{30}{4.5} = 6.67 < 10,\ c_1 = 2.012,$

$$\frac{h_2}{b_2} = \frac{60}{2.5} = 24 > 10,\quad \frac{h_3}{b_3} = \frac{40}{5.5} = 7.27 < 10,\ c_1 = 2.212$$

$$I_{t_1} = c_1 b_1^4 = 2.012\times4.5^4 = 825.04\ \text{cm}^4,\quad I_{t_3} = c_1 b_3^4 = 2.212\times5.5^4 = 2024.12\ \text{cm}^4$$

$$I_{t_2} = \frac{h_2 b_2^3}{3} = \frac{60\times2.5^3}{3} = 312.5\ \text{cm}^4,\quad \sum I_t = I_{t_1} + I_{t_2} + I_{t_3} = 3161.66\ \text{cm}^4$$

$$S_t = \frac{I_t}{b_{max}} = \frac{3161.66}{5.5} = 574.85\ \text{cm}^3,$$

$$\tau_{max} = \frac{40\times(100)}{574.85} = 6.958\ \text{kN/cm}^2 = 69580\ \text{kN/m}^2 = 69.58\ \text{MPa}$$

$$\varphi^0 = \frac{180}{\pi}\cdot\frac{M_t L}{G I_t} = \frac{180}{3.14}\cdot\frac{40\times(100)\times1.5\times(100)}{800\times3161.66} = 13.6^0$$

STRESS and STRAIN

TORSION

1.5

Bar of circular cross-section

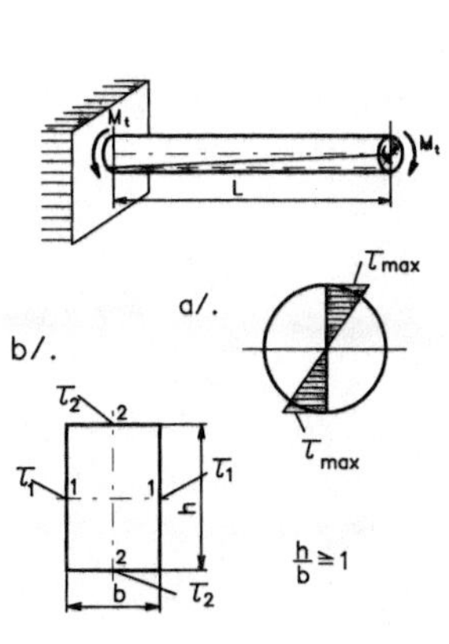

Stress: $\tau_{max} = \dfrac{M_t}{I_p} \cdot \dfrac{d}{2} = \dfrac{M_t}{S_p}$,

$I_p = \dfrac{\pi d^4}{32} \approx 0.1 d^4$, $S_p = \dfrac{\pi d^3}{16} \approx 0.2 d^3$.

Angle of twist: $\varphi^0 = \dfrac{180^0}{\pi} \cdot \dfrac{M_t L}{G I_p}$.

Where G = Shear modulus of elasticity

Bar of rectangular cross-section

Stress: $\tau_{max} \dfrac{M_t}{S_t}$. Angle of twist: $\varphi^0 = \dfrac{180^0}{\pi} \cdot \dfrac{M_t L}{G I_t}$.

If $\dfrac{h}{b} > 10$: $I_t = \dfrac{hb^3}{3}$, $S_t = \dfrac{I_t}{b} = \dfrac{hb^2}{3}$.

If $\dfrac{h}{b} \le 10$: $I_t = c_1 \cdot b^4$, $S_t = c_2 \cdot b^3$.

In point 1: $\tau_1 = \tau_{max}$, in point 2: $\tau_2 = c_3 \cdot \tau_{max}$.

h / b =	1.0	1.5	2.0	3.0	4.0	6.0	8.0	10.0	For
c_1	0.140	0.294	0.457	0.790	1.123	1.789	2.456	3.123	h / b >10
c_2	0.208	0.346	0.493	0.801	1.150	1.789	2.456	3.123	
c_3	1.000	0.859	0.795	0.753	0.745	0.743	0.742	0.742	0.740

Profile consisting of rectangular cross-sections

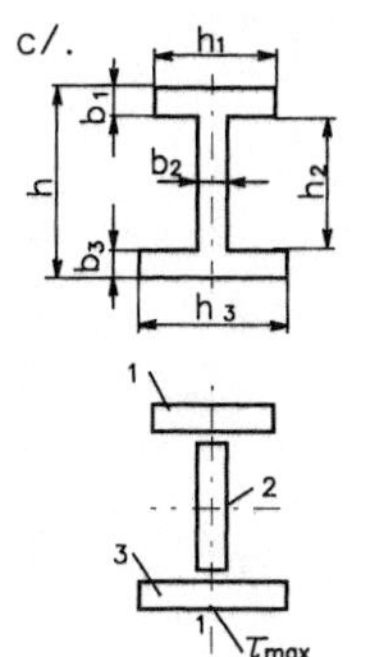

Geometric properties: $I_t = \sum_{i=1}^{i=n} I_{t_i}$, $S_t = \dfrac{I_t}{b_{max}}$, $n = 3$

Assumed: $\dfrac{h_1}{b_1} < 10$, $\dfrac{h_2}{b_2} > 10$, $\dfrac{h_3}{b_3} < 10$,

$b_3 > b_1$, $b_3 > b_2$ (i.e. $b_3 = b_{max}$)

$I_{t_1} = c_1 b_1^4$, $I_{t_2} = \dfrac{h_2 b_2^3}{3}$, $I_{t_3} = c_1 b_3^4$,

$I_t = I_{t_1} + I_{t_2} + I_{t_3}$, $S_t = \dfrac{I_t}{b_3}$.

Stress: $\tau_{max} = \dfrac{M_t}{S_t}$ (in point 1).

Angle of twist: $\varphi^0 = \dfrac{180^0}{\pi} \cdot \dfrac{M_t L}{G I_t}$.

NOTES

Curved beam (transverse bending)

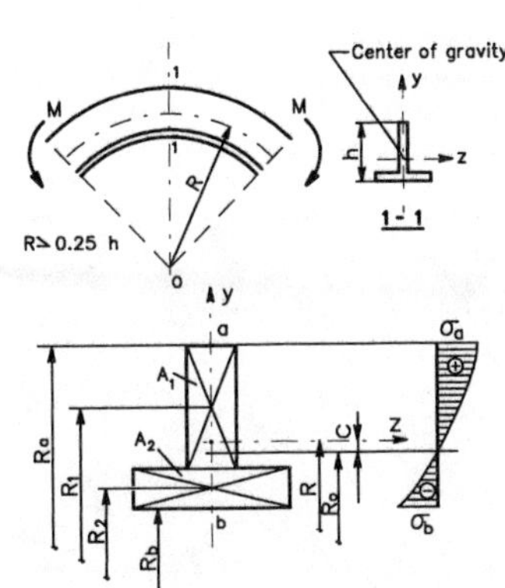

Stresses:

$$\sigma_y = \frac{M}{A \cdot c} \cdot \frac{y - R_0}{y}, \qquad R_0 = \frac{\sum A_i}{\sum \frac{A_i}{R_i}}.$$

$$c = R - R_0$$

If $\frac{h}{R} \leq 0.5$, $\quad c = \frac{I_z}{A \cdot R}$ for all cross-section types.

For case shown:

$$A = A_1 + A_2, \qquad R_0 = \frac{A_1 + A_2}{\frac{A_1}{R_1} + \frac{A_2}{R_2}},$$

$$\sigma_a = \frac{M}{A \cdot c} \cdot \frac{R_a - R_0}{R_a}, \qquad \sigma_b = \frac{M}{A \cdot c} \cdot \frac{R_b - R_0}{R_b}.$$

"$+\sigma$ " - Tension

"$-\sigma$ " - Compression

Curved beam (axial force and bending)

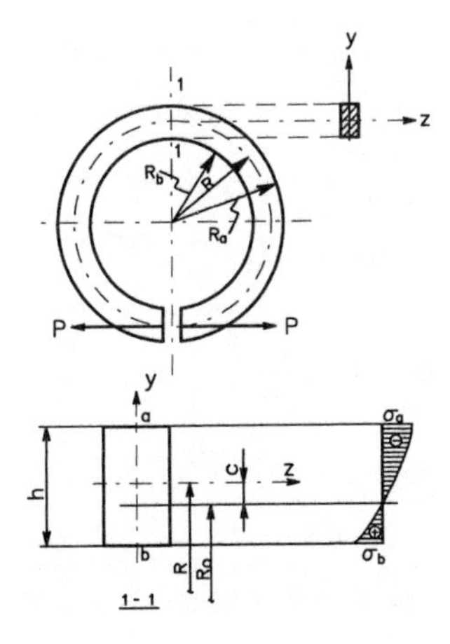

Stresses: $\quad \sigma_\rho = \frac{N}{A} \pm \frac{M}{A \cdot c} \cdot \frac{\rho - R_0}{R_0}.$

For case shown: $\quad c = R - R_0,$

$$R_0 = \frac{h}{\ln \frac{R_a}{R_b}} \quad \text{or} \quad R_0 \approx R\left[1 - \frac{1}{12}\left(\frac{h}{R}\right)^2\right].$$

$$N = P, \quad M = 2PR,$$

$$\sigma_a = \frac{P}{bh} - \frac{2PR}{bhc} \cdot \frac{R_a - R_0}{R_a},$$

$$\sigma_b = \frac{P}{bh} + \frac{2PR}{bhc} \cdot \frac{R_0 - R_b}{R_b}.$$

Note. For beams with circular cross-section:

$$R_0 = \frac{1}{2}\left(R + \sqrt{R^2 - \frac{d^2}{R}}\right) \quad \text{or} \quad R_0 \approx R\left[1 - \frac{1}{16}\left(\frac{d}{R}\right)^2\right],$$

d = diameter of cross-section.

N O T E S

Table 1.7

Example. Continuous deep beam

Given. Beam $L = 3.0$ m, $h = 2.0$ m, $c = 0.3$ m, thickness $b = 0.3$ m, $w = 200$ kN/m

Required. Compute Z, D, d, d_0 and σ_{max} for center of span and support

Solution. At center of span:

$$Z = D = \alpha_z \times 0.5wL = 0.186 \times 0.5 \times 200 \times 3.0 = 55.8 \text{ kN}$$

$$d = \alpha_d \times 0.5L = 0.888 \times 0.5 \times 3.0 = 1.33 \text{ m}$$

$$d_0 = \alpha_{d_0} \times 0.5L = 0.124 \times 0.5 \times 3.0 = 0.19 \text{ m}$$

$$\sigma_{max} = \alpha_\sigma \times w / b = 1.065 \times 200 / 0.3 = 710 \text{ kN/m}^2 = 0.71 \text{ MPa} \quad \text{(tension)}$$

At center of support:

$$Z = D = \alpha_z \times 0.5wL = 0.428 \times 0.5 \times 200 \times 3.0 = 128.4 \text{ kN}$$

$$d = \alpha_d \times 0.5L = 0.656 \times 0.5 \times 3.0 = 0.984 \text{ m}$$

$$d_0 = \alpha_{d_0} \times 0.5L = 0.036 \times 0.5 \times 3.0 = 0.05 \text{ m}$$

$$\sigma_{max} = \alpha_\sigma \times w / b = -9.065 \times 200 / 0.3 = -6043.3 \text{ kN/m}^2 = -6.04 \text{ MPa} \quad \text{(compression)}$$

$h \geq 0.5L$

Stress diagrams

Formulas: Maximum tensile and compressive stresses $\sigma_{max} = \alpha_\sigma \cdot w$

Resultant tensile (Z) and compressive (D) forces $Z = D = \alpha_z \cdot 0.5wL$

$d = \alpha_d \cdot 0.5L$, $d_0 = \alpha_{d(0)} \cdot 0.5L$,

Coefficients

h/L	α	At center of span			At center of support		
		c/L			c/L		
		0.05	0.10	0.20	0.05	0.10	0.20
0.5	α_σ	1.317	1.313	1.289	–19.320	–9.317	–4.302
	α_z	0.240	0.239	0.235	0.515	0.485	0.375
	α_d	0.692	0.690	0.682	0.600	0.622	0.640
	$\alpha_{d(0)}$	0.129	0.128	0.127	0.022	0.039	0.062
0.67	α_σ	1.066	1.065	1.062	–19.066	–9.065	–4.062
	α_z	0.187	0.186	0.182	0.498	0.428	0.351
	α_d	0.890	0.888	0.880	0.620	0.656	0.686
	$\alpha_{d(0)}$	0.125	0.124	0.122	0.021	0.036	0.059
1.0	α_σ	1.002	1.002	1.002	–19.002	–9.002	–4.002
	α_z	0.178	0.177	0.172	0.497	0.424	0.324
	α_d	0.934	0.932	0.924	0.612	0.682	0.740
	$\alpha_{d(0)}$	0.124	0.123	0.121	0.021	0.036	0.059
$h = \infty$	α_σ	1.000	1.000	1.000	–19.000	–9.000	–4.000
	α_z	0.177	0.176	0.171	0.495	0.422	0.322
	α_d	0.938	0.936	0.930	0.612	0.674	0.746
	$\alpha_{d(0)}$	0.122	0.122	0.121	0.024	0.038	0.059

NOTES

Tables 1.8–1.12 consider computation methods for elastic systems only.

STRESS and STRAIN

DYNAMICS, TRANSVERSE OSCILLATIONS OF THE BEAMS 1.8

NATURAL OSCILLATIONS OF SYSTEMS WITH ONE DEGREE FREEDOM

1 SIMPLE BEAM WITH ONE POINT MASS

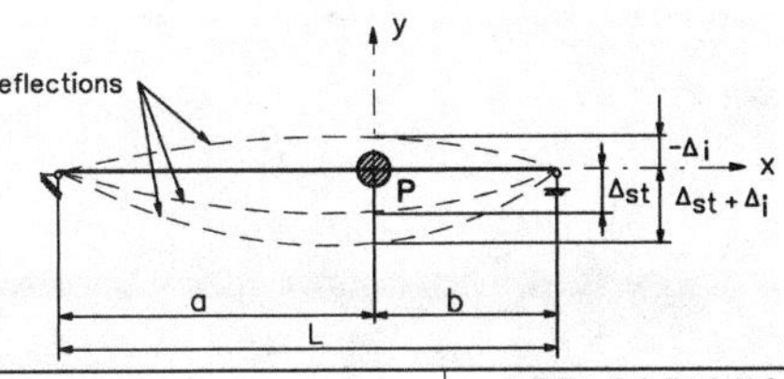

FORCES:

P = Weight of the load, Mass: $m = \frac{P}{g}$

g = Gravitational acceleration, $\left(g = 981\frac{cm}{sec^2}\right)$

P_i = Force of inertia, $P_i = \mp ma$

a = acceleration

For shown beam:

Maximum Bending Moment

$M_{max} = (P + P_i)\cdot\frac{a\cdot b}{L}$, Stress: $\sigma = \frac{M_{max}}{I_z}\cdot y$

DEFLECTIONS:

Δ_{st} = Static deflection due to Load P

$\pm\Delta_i$ = Max., min. deflection due to Force P_i

$\Delta_{st(1)}$ = Static deflection due to Force $P = 1$

c = amplitude, $c = \pm\Delta_i$

Maximum Shear for $a > b$

$V_{max} = (P + P_i)\cdot\frac{a}{L}$

Stress: $\tau = \frac{V_{max}\cdot S}{I_z \cdot t}$

2

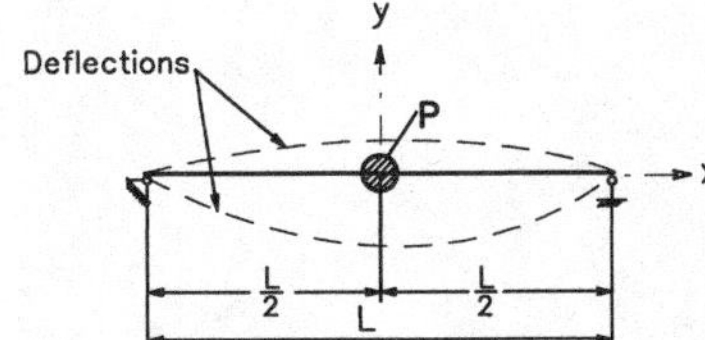

Force of inertia: $P_i = \frac{48cEI_z}{L^3}$

Maximum Bending Moment: $M_{max} = \left(\frac{48cEI_z}{L^3} + P\right)\cdot\frac{L}{4}$

Maximum Shear: $V_{max} = \frac{1}{2}\left(\frac{48cEI_z}{L^3} + P\right)$

3

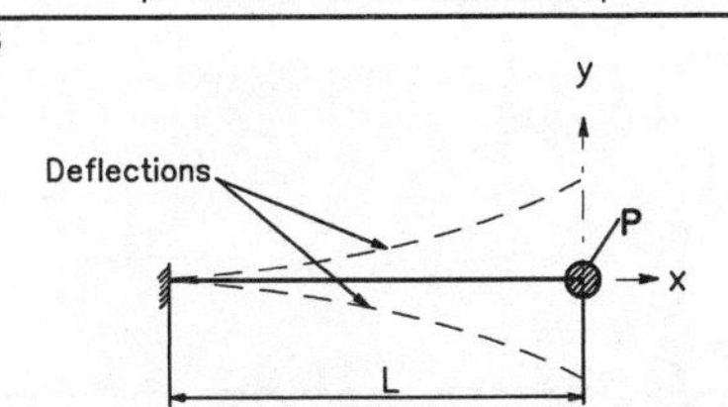

Force of inertia: $P_i = \frac{3cEI_z}{L^3}$

Maximum Bending Moment: $M_{max} = \left(\frac{3cEI_z}{L^3} + P\right)\cdot L$

Maximum Shear: $V_{max} = \frac{3cEI_z}{L^3} + P$

NOTES

STRESS and STRAIN

DYNAMICS, TRANSVERSE OSCILLATIONS OF THE BEAMS — 1.9

DIAGRAM OF CONTINUOUS OSCILLATIONS

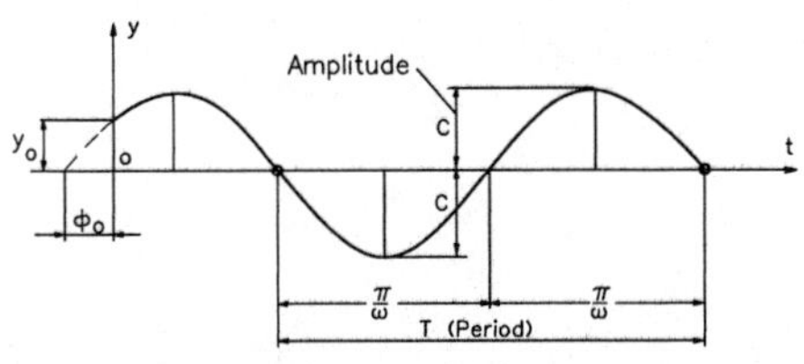

Equation of free continuous oscillations: $y = c\sin(\omega t + \varphi_0)$

Where: ϕ_0 = initial phase of oscillation, $\phi_0 = \arcsin\left(\frac{y_0}{c}\right)$

c_0 = amplitude, t = time, T = period of free oscillation, $T = \frac{2\pi}{\omega} = 2\pi\sqrt{\frac{\Delta_{st}}{g}}$

ω = frequency of natural oscillation, $\omega = \sqrt{\frac{g}{\Delta_{st}}}$

DIAGRAM OF DAMPED OSCILLATIONS

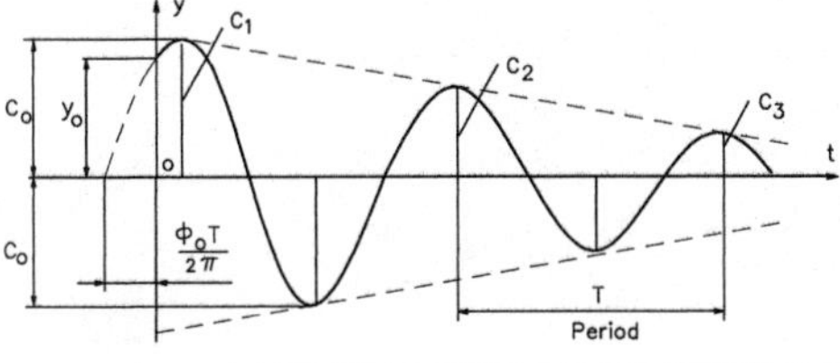

Equation of free damped oscillations: $y = c_0 e^{-kt/2m} \cdot \sin(\omega t + \varphi_0)$

c_0 = initial amplitude of oscillation, $c_0 = \sqrt{y_0^2 + \left(\frac{v_0 + y_0 k \cdot 2m}{\omega}\right)^2}$

ϕ_0 = initial phase of oscillation, $\phi_0 = \arcsin\left(\frac{y_0}{c_0}\right)$, y_0 = initial deflection

v_0 = beginner velocity of mass, e = logarithmic base, $e = 2.71828$

k = coefficient set according to material, mass and rigidity

T = period of free oscillations, $T = 2\pi/\omega$

ω = frequency of free oscillation, $\omega = \sqrt{r/m - [k/2m]^2}$, For simple beam: $r = \frac{48EI_z}{L^3}$

N O T E S

FORCED OSCILLATIONS OF THE BEAMS WITH ONE DEGREE FREEDOM

SIMPLE BEAM WITH ONE POINT MASS

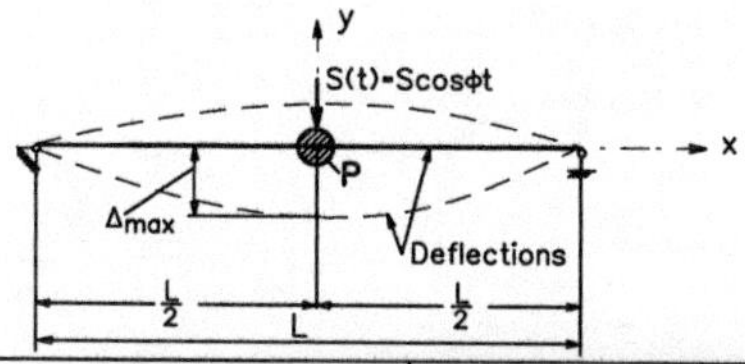

FORCES:	DEFLECTIONS:
$P =$ Weight of the load, Mass: $m = \frac{P}{g}$, $\left(g = 981 \frac{cm}{sec^2}\right)$	$\Delta_{max} = \Delta_{st(p)} + \Delta_{st(s)} + \Delta_i$
$S(t) =$ vibrating force, Assumed: $S(t) = S\cos\varphi t$	$\Delta_{st(p)} =$ Static deflection due to Load P
$P_i =$ Force of inertia, $P_i = \frac{\Delta_{max} - \Delta_{st}}{\Delta st_1} - S\cos\varphi t$	$\Delta_{st(s)} =$ Static deflection due to Force S
$\varphi =$ Frequency of force $S(t)$	$\Delta_i =$ Static deflection due to P_i,
$\Delta_{st(1)} =$ Static deflection due to Load $P = 1$	$\Delta_i = P_i \cdot \Delta_{st(1)}$

Equation of forced oscillations: $y = c \cdot e^{-kt/2m} \cdot \sin(\omega t + \varphi_0) + \frac{g \cdot S(t)}{P(\omega^2 - \varphi^2)} \cdot \cos\varphi t$

$c \cdot e^{-kt/2m} \cdot \sin(\omega t + \varphi_0) =$ free oscillation, $\frac{g \cdot S(t)}{P(\omega^2 - \varphi^2)} \cdot \cos\varphi t =$ forced oscillation

$\varphi_0 =$ beginner phase of oscillation, $\varphi_0 = \arcsin\left(\frac{y_0}{c_0}\right)$, $y_0 =$ beginner deflection

$c_0 =$ amplitude of free oscillation, $c_0 = c$, $c =$ amplitude of forced oscillation, $c = k_D \cdot \Delta_{st(s)}$

$k =$ coefficient set according to material, mass and rigidity

$\omega =$ frequency of natural oscillation, $T =$ period of oscillations, $T = 2\pi/\omega$

$k_D =$ dynamic coefficient, $k_D = \frac{1}{\sqrt{\left(1 - \frac{\varphi^2}{\omega^2}\right)^2 + \left[\frac{k \cdot \varphi}{m \cdot \omega^2}\right]^2}}$

If $k = 0$ (damped oscillation is not included): $k_D = \frac{1}{1 - \frac{\varphi^2}{\omega^2}}$

$e =$ logarithmic base, $e = 2.71828$, $g =$ gravitational acceleration, $\left(g = 981 \frac{cm}{sec^2}\right)$

N O T E S

Table 1.11 Dynamics, impact

Example. Bending

Given. Beam $W12\times 65$, Steel, $L = 3.0$ m,

Moment of inertia $I_z = 533\ \text{in}^4 \times 2.54^4 = 22185\ \text{cm}^4$

Section modulus $S = 87.9\ \text{in}^3 = 87.9\times 2.54^3 = 1440.4\ \text{cm}^3$

Modulus of elasticity $E = 29000\ \text{kip/in}^2 = \dfrac{29000\times 4.48222}{2.54^2} = 20147.6\ \text{kN/cm}^2$

Weight of beam (concentrated load):

$W = 65\ \text{Lb/ft}\times 3.0 = 195\times 4.448/0.3048 = 2845.7\ \text{N} = 2.8457\ \text{kN}$

Load $P = 20$ kN, $h = 5$ cm

Required. Compute dynamic stress σ

Solution.
$$\Delta_{st} = \frac{PL^3}{48EI_z} = \frac{20\times(3\times 100)^3}{48\times 20147.6\times 22185} = 0.025\ \text{cm}$$

$$k_D = 1 + \sqrt{1 + \frac{2h}{\Delta_{st}\left(1+\beta\dfrac{W}{P}\right)}} = 1 + \sqrt{1 + \frac{2\times 5}{0.025\left(1+\dfrac{17}{35}\times\dfrac{2.8457}{20}\right)}} = 1 + 19.4 = 20.4$$

Bending moment $M_D = \dfrac{PL}{4}\cdot k_D = \dfrac{20\times 3}{4}\times 20.4 = 306\ \text{kN}\cdot\text{m}$

Stress $\sigma = \dfrac{M_D}{S} = \dfrac{306\times 100}{1440.4} = 21.24\ \text{kN/cm}^2 = 212400\ \text{kN/m}^2 = 212.4\ \text{MPa}$

Table 1.11 Dynamics, impact

Example. Crane cable

Given. Load $P = 40$ kN, velocity $\upsilon = 5$ m/sec

Cable: diameter $d = 5.0$ cm, $A = 19.625\ \text{cm}^2$, $L = 30$ m,

Modulus of elasticity $E = 29000\ \text{kip/in}^2 = \dfrac{29000\times 4.48222}{2.54^2} = 20147.6\ \text{kN/cm}^2$

Required. Compute dynamic stress σ for sudden dead stop

Solution.
$$\Delta_{st} = \frac{PL}{EA} = \frac{40\times 30\times(100)}{20147.6\times 19.625} = 0.303\ \text{cm},\quad k_D = \frac{\upsilon}{\sqrt{g\cdot\Delta_{st}}} = \frac{5\times(100)}{\sqrt{981\times(100)\times 0.303}} = 2.9$$

Stress:
$$\sigma = \frac{P}{A}(1+k_D) = \frac{40}{19.625}(1+2.9) = 7.949\ \text{kN/cm}^2 = 79490\ \text{kN/m}^2 = 79.45\ \text{MPa}$$

Elastic design

Axial compression

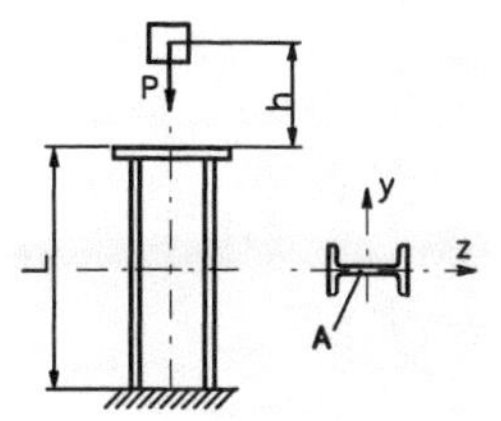

Dynamic coefficient:

$$k_D = 1 + \sqrt{1 + \frac{\upsilon^2}{g\Delta_{st}\left(1 + \beta\frac{W}{P}\right)}} = 1 + \sqrt{1 + \frac{2h}{\Delta_{st}\left(1 + \beta\frac{W}{P}\right)}}$$

Where:

υ= strking velocity, $\upsilon = \sqrt{2gh}$

g= earth's acceleration , $g = 9.81\ m/sec^2$

Δ_{st}= deflection resulting from static load P

W= weight of the structure

β= coefficient for uniform mass

For shown column: $\Delta_{st} = \frac{PL}{EA}$, $\beta = \frac{1}{3}$.

Dynamic stress : $\sigma = -\frac{P}{A}\cdot k_D$,

Bending

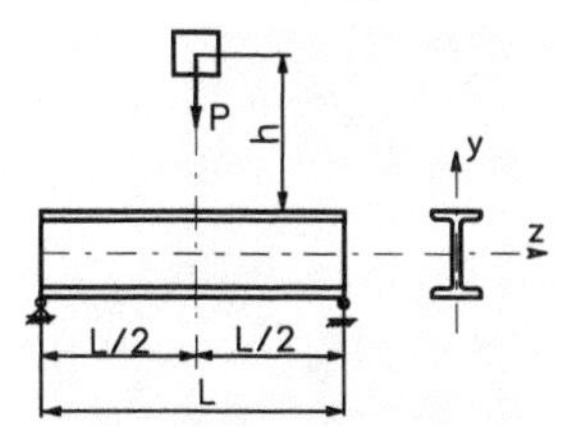

For shown beam: $\Delta_{st} = \frac{PL^3}{48EI_z}$, $\beta = \frac{17}{35}$.

Dynamic bending moment: $M_D = \frac{PL}{4}\cdot k_D$,

Dynamic shear: $V_D = \frac{P}{2}\cdot k_D$.

For stresses see Table 1.3

Crane cable

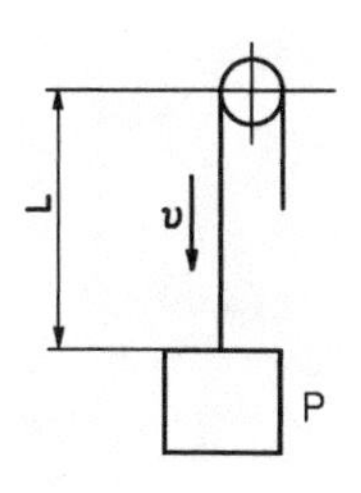

Sudden dead stop when the load P is going down.

Dynamic coefficient:

$$k_D = \frac{\upsilon}{\sqrt{g\cdot\Delta_{st}}} ,$$

where: υ = descent's velocity ,

$$\Delta_{st} = \frac{PL}{EA} .$$

Maximum stress in the cable:

$$\sigma = \frac{P}{A}(1 + k_D)$$

A = area of cable cross-section

NOTES

Elastic design

Column with buffer spring

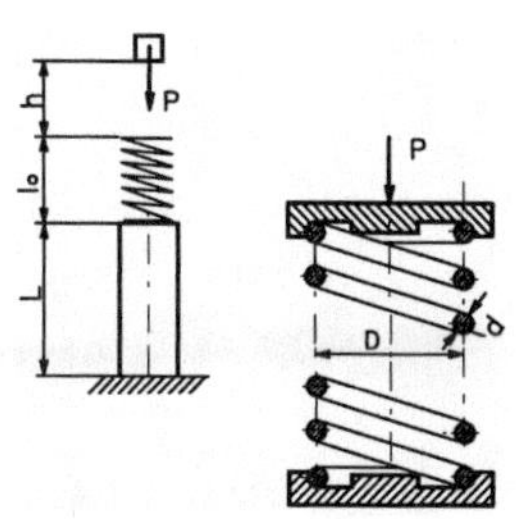

Cylindrical helical spring:

D = average diameter

d = spring wire's diameter

n = number of effective rings

G= Shear modulus of elasticity for spring wire

Dynamic coefficient:

$$k_D = 1 + \sqrt{1 + \frac{2h}{P\left(\frac{8D^3 n}{Gd^4} + \frac{L}{EA}\right)}}\ .$$

Dynamic stress: $\sigma = -\frac{P}{A} \cdot k_D$ (compression)

E= Modulus of elasticity for column

A= area of column cross-section

Motor mounted on the beam

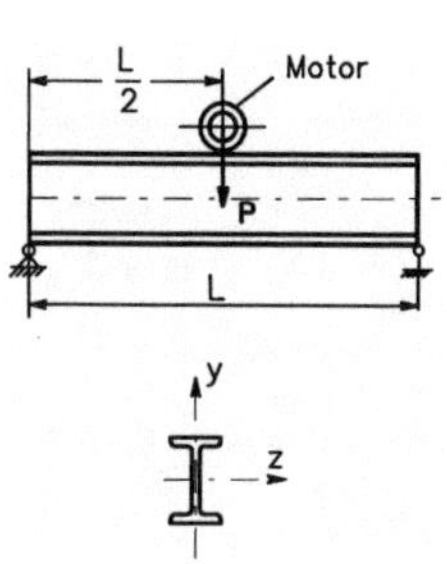

P = motor's weight,

F_c = centrifugal force causing vertical vibration of the beam, $F_c = m\varphi^2 r$,

m = mass of rotative motor part ,

r = radius of rotation ,

n = revolutions per minute.

Dynamic coefficient: $k_D = \frac{1}{1 - \frac{\varphi^2}{\omega^2}}$,

φ = frequency of force F_c, $\varphi = \frac{n}{60} \cdot 2\pi = \frac{\pi n}{30} \left(\frac{1}{\text{sec}}\right)$

ω = beam's free vibration frequency, $\omega = \sqrt{\frac{g}{P\Delta}} \left(\frac{1}{\text{sec}}\right)$

Δ = beam's deflection by force $P = 1$ at the point of motor attachment,

$\left(\text{For shown case: } \Delta = \frac{L^3}{48EI_z}\right)$.

Resonance: $\varphi = \omega$, $n = \frac{30\varphi}{\pi}$.

Stresses:

Static stress: $\sigma = \frac{PL}{4S_z}$, Dynamic stress: $\sigma = \frac{F_c k_D L}{4S_z}$,

$$\sum \sigma = \frac{L}{4S_z}(P + F_c k_D)$$

NOTES

2. PROPERTIES OF GEOMETRIC SECTIONS

NOTES

PROPERTIES OF GEOMETRIC SECTIONS

for TENSION, COMPRESSION, and BENDING STRUCTURES **2.1**

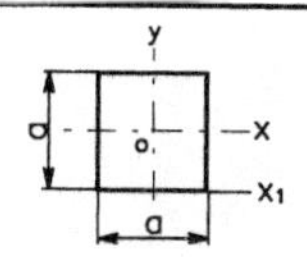

1. SQUARE

$$A = a^2, \quad I_x = I_y = \frac{a^4}{12}, \quad I_{x_1} = \frac{a^4}{3},$$

$$S_x = S_y = \frac{a^2}{6}, \quad r_x = r_y = \frac{a}{\sqrt{12}} = 0.289a \ , \ Z = \frac{a^3}{4}$$

2. SQUARE

Axis of moments on diagonal

$$A = a^2, \quad h = a\sqrt{2} = 1.42a, \ I_x = I_y = \frac{a^4}{12}, \quad S_x = S_y = \frac{a^3}{6\sqrt{2}} = 0.118a^3,$$

$$r_x = r_y = \frac{a}{\sqrt{12}} = 0.289a, \quad Z = \frac{a}{3\sqrt{2}} = 0.236a$$

3. RECTANGLE

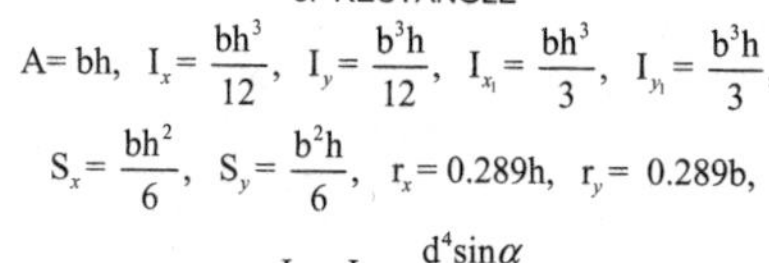

$$A = bh, \quad I_x = \frac{bh^3}{12}, \quad I_y = \frac{b^3h}{12}, \quad I_{x_1} = \frac{bh^3}{3}, \quad I_{y_1} = \frac{b^3h}{3},$$

$$S_x = \frac{bh^2}{6}, \quad S_y = \frac{b^2h}{6}, \quad r_x = 0.289h, \quad r_y = 0.289b,$$

$$I_{x_2} = I_{x_3} = \frac{d^4 \sin\alpha}{48}$$

4. RECTANGLE

Axis of moments on any line through center of gravity

$$A = bh, \quad y_t = y_b = \frac{1}{2}(h\cos\alpha + b\sin\alpha),$$

$$I_x = \frac{bh}{12}(h^2\cos^2\alpha + b^2\sin^2\alpha), \qquad S_x = \frac{bh\,(h^2\cos^2\alpha + b^2\sin^2\alpha)}{6(h\cos\alpha + b\sin\alpha)},$$

$$r_x = 0.289\sqrt{(h^2\cos^2 + b^2\sin^2\alpha)}$$

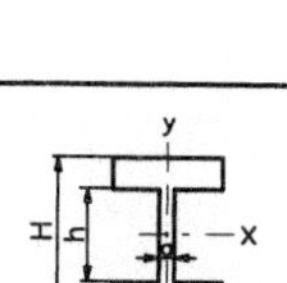

5. SYMMETRICAL SHAPE

$$A = ah + b(H - h),$$

$$I_x = \frac{ah^3}{12} + \frac{b}{12}(H^3 - h^3), \quad I_y = \frac{a^3h}{12} + \frac{b^3}{12}(H - h),$$

$$S_x = \frac{b}{6H}(H^3 - h^3) + \frac{ah^3}{6H}, \quad S_y = \frac{a^3h}{6b} + \frac{b^2}{6}(H - h)$$

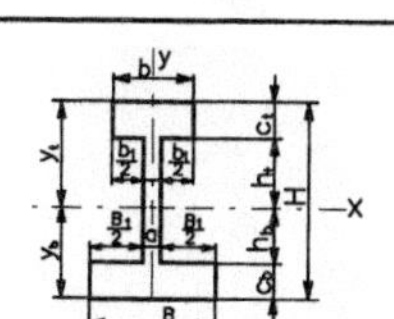

6. NONSYMMETRICAL SHAPE

$$A = bc_t + a(h_b + h_t) + Bc_b, \quad b_1 = b - a, \quad B_1 = B - a,$$

$$y_b = \frac{aH^2 + B_1c_b + b_1c_t(2H - c_t)}{2(aH + B_1c_b + b_1c_t)}, \quad y_t = H - y_b \ ,$$

$$I_x = \frac{1}{3}(By_b^3 - B_1h_b^3 + by_t^3 - b_1h_t^3).$$

NOTES

PROPERTIES OF GEOMETRIC SECTIONS

for TENSION, COMPRESSION, and BENDING STRUCTURES

2.2

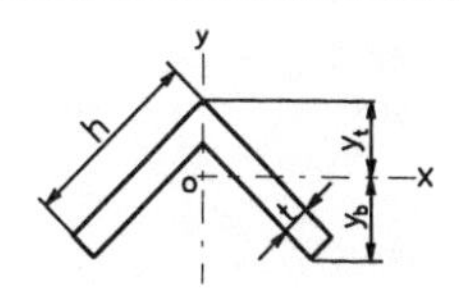

7. ANGLE with equal legs

$$A = t(2h - t), \quad y_t = \frac{h^2 + ht + t^2}{2(2h - t)\cos 45^0}, \quad y_b = \frac{h + t - 2c}{\sqrt{2}},$$

$$I_x = \frac{1}{3}\left[2c^4 - 2(c - t)^4 + t(h - 2c + \frac{1}{2}t)^3\right]$$

$$c = y_t \cos 45^0$$

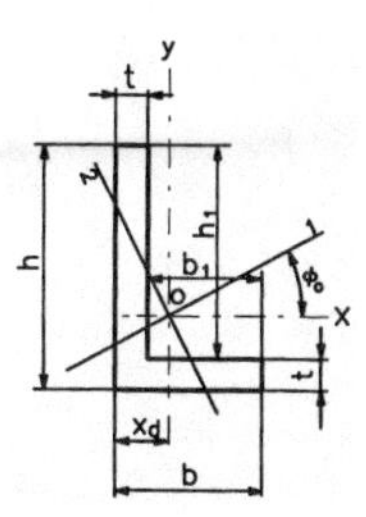

8. ANGLE with unequal legs

$$A = t(b + h_1) = t(h + b_1), \quad x_d = \frac{b^2 + h_1 t}{2(b + h_1)}, \quad y_d = \frac{h^2 + b_1 t}{2(h + b_1)},$$

$$I_x = \frac{1}{3}\left[t\left(h - y_d\right)^3 + by_d^3 - b_1\left(y_d - t\right)^3\right],$$

$$I_y = \frac{1}{3}\left[t\left(b - x_d\right)^3 + hx_d^3 - h_1\left(x_d - t\right)^3\right]$$

$$I_1 = I_{max} \text{ and } I_2 = I_{min}, \quad \tan 2\varphi_0 = \frac{2I_{xy}}{I_y - I_x},$$

I_{xy} = Product of inertia about axes x and y, $I_{xy} = \pm\dfrac{bb_1hh_1t}{4(b+h_1)}$,

$$I_{1(2)} = I_{max(min)} = \frac{1}{2}\left(I_y + I_x\right) \pm \frac{1}{2}\sqrt{\left(I_y - I_x\right)^2 + 4I_{xy}^2},$$

9. TRIANGLE

$$A = \frac{1}{2}bh, \quad h_b = \frac{1}{3}h, \quad h_t = \frac{2}{3}h, \quad d = \frac{1}{3}(b_a - b_c),$$

$$I_x = \frac{bh^3}{36}, \quad I_{x_1} = \frac{bh^3}{12}, \quad I_{x_2} = \frac{bh^3}{4},$$

$$I_y = \frac{hb\left(b^2 - b_a b_c\right)}{36}, \quad I_{y_1} = \frac{h\left(b_a^3 + b_c^3\right)}{12},$$

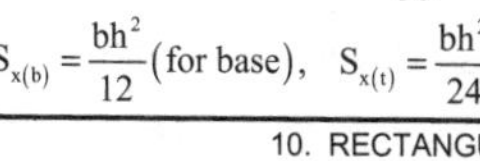

$$S_{x(b)} = \frac{bh^2}{12}(\text{for base}), \quad S_{x(t)} = \frac{bh^2}{24}(\text{for point A}), \quad r_x = \frac{h}{3\sqrt{2}} = 0.236h.$$

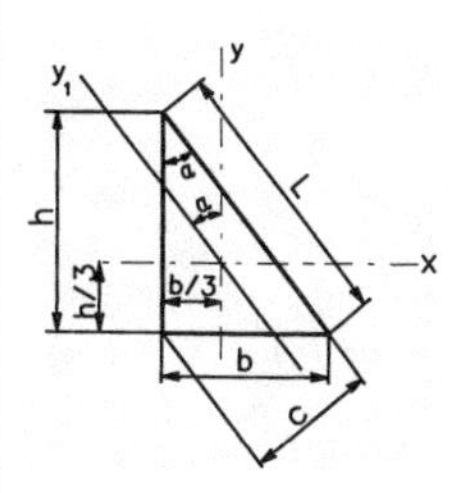

10. RECTANGULAR TRIANGLE

$$A = \frac{bh}{2} = \frac{cL}{2}, \quad I_x = \frac{bh^3}{36}, \quad I_y = \frac{hb^3}{36},$$

$$I_{y_1} = \frac{b^3h^3}{36\,L^2} = \frac{Lc^3}{36},$$

$$\text{or: } I_{y_1} = I_y\cos^2\alpha + I_x\sin^2\alpha + 2I_{xy}\sin\alpha\cos\alpha,$$

$$\sin\alpha = \frac{b}{L}, \quad \cos\alpha = \frac{h}{L}, \quad I_{xy} = -\frac{b^2h^2}{72},$$

$$r_x = \frac{h}{3\sqrt{2}} = 0.236h.$$

NOTES

PROPERTIES OF GEOMETRIC SECTIONS

for TENSION, COMPRESSION, and BENDING STRUCTURES

2.3

11. TRAPEZOID

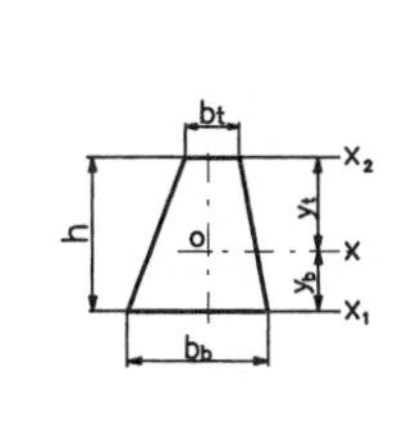

$$A=\frac{1}{2}\left(b_t+b_b\right)h,\quad y_b=\frac{b_b+2b_t}{3\left(b_b+b_t\right)}h,\quad y_t=\frac{2b_b+b_t}{3\left(b_b+b_t\right)}h,$$

$$I_x=\frac{h^3\left(b_b^2+4b_bb_t+b_t^2\right)}{36\left(b_b+b_t\right)},\quad I_{x_1}=\frac{h^3\left(b_b+3b_t\right)}{12},$$

$$I_{x_2}=\frac{h^3\left(3b_b+b_t\right)}{12},\quad S_{x_b}=\frac{I_x}{y_b}(\text{bottom}),\quad S_{x_t}=\frac{I_x}{y_t}(\text{top}),$$

$$r_x=\frac{h\sqrt{2\left(b_b^2+4b_bb_t+b_t^2\right)}}{6\left(b_b+b_t\right)}.$$

12. REGULAR HEXAGON

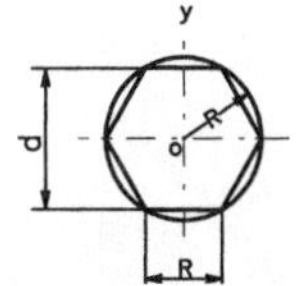

$$A=2.598R^2=0.866d^2,\quad I_x=I_y=0.541R^4=0.06d^4,$$

$$S_x=0.625R^3,\quad S_y=0.541R^3,$$

$$r_x=r_y=0.456R=0.263d.$$

13. REGULAR OCTAGON

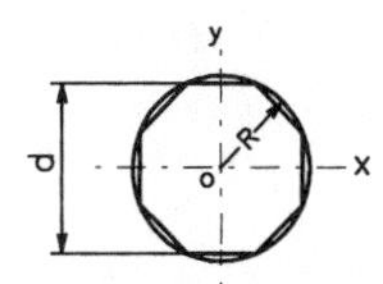

$$A=0.828d^2,\quad I_x=I_y=0.638R^4=0.0547d^4,$$

$$S_x=S_y=0.690R^3=0.1095d^3,\quad r_x=r_y=0.257d.$$

14. REGULAR POLYGON with n sides

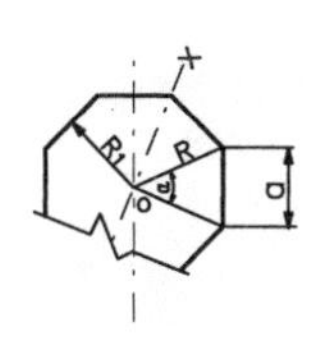

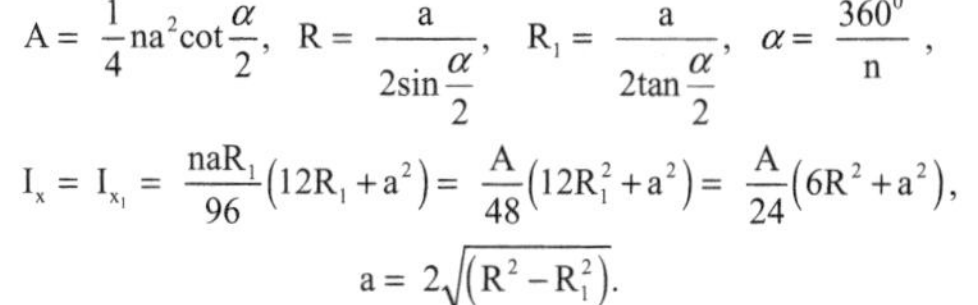

$$A=\frac{1}{4}na^2\cot\frac{\alpha}{2},\quad R=\frac{a}{2\sin\frac{\alpha}{2}},\quad R_1=\frac{a}{2\tan\frac{\alpha}{2}},\quad \alpha=\frac{360^0}{n},$$

$$I_x=I_{x_1}=\frac{naR_1}{96}\left(12R_1+a^2\right)=\frac{A}{48}\left(12R_1^2+a^2\right)=\frac{A}{24}\left(6R^2+a^2\right),$$

$$a=2\sqrt{\left(R^2-R_1^2\right)}.$$

15. CIRCLE

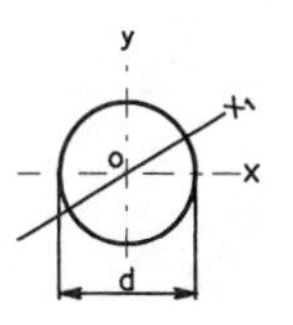

$$A=\frac{\pi d^2}{4}\approx 0.785d^2,\quad I_x=I_y=I_{x_1}=\frac{\pi d^4}{64}\approx 0.05d^4,$$

$$S_x=S_y=S_{x_1}=\frac{\pi d^3}{32}\approx 0.1d^3,$$

$$r_x=r_y=\frac{d}{4},\quad Z=\frac{d^3}{6}.$$

N O T E S

PROPERTIES OF GEOMETRIC SECTIONS

for TENSION, COMPRESSION, and BENDING STRUCTURES **2.4**

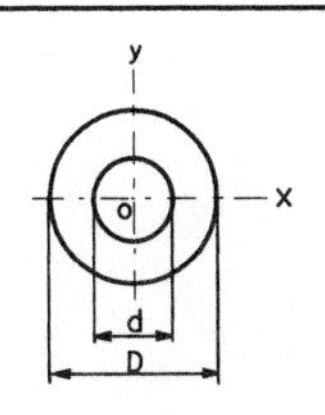

16. HOLLOW CIRCLE

$$A=\frac{\pi D^2}{4}\left(1-\xi^2\right),\quad \xi=\frac{d}{D},\quad I_x=I_y=\frac{\pi D^4}{64}\left(1-\xi^4\right),$$

$$S_x=S_y=\frac{\pi D^3}{32}\left(1-\xi^4\right),\quad r_x=r_y=\frac{D}{4}\sqrt{1-\xi^2},$$

$$Z=\frac{D^3-d^3}{6}.$$

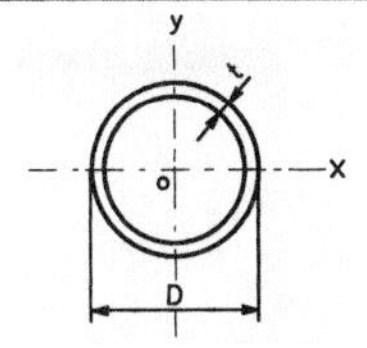

17. THIN RING (t<<D)

$$A=\pi Dt,\quad I_x=\frac{\pi D^3 t}{8}\approx 0.3926D^3t,$$

$$S_x=\frac{\pi D^2 t}{4}\approx 0.7853D^2t,\quad r_x=0.353D.$$

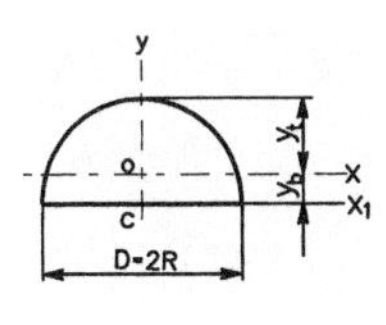

18. Half of a CIRCLE

$$A=\frac{\pi D^2}{8}\approx 0.392D^2,\quad y_b=0.2122D,\quad y_t=0.2878D,$$

$$I_x=0.00686D^4,\quad I_y=I_{x_1}=\frac{\pi D^4}{128}\approx 0.025D^4,$$

$$S_{x_b}=0.2587\left(\frac{D}{2}\right)^3-\text{for bottom},\quad S_{x_t}=0.1908\left(\frac{D}{2}\right)^3-\text{for top}.$$

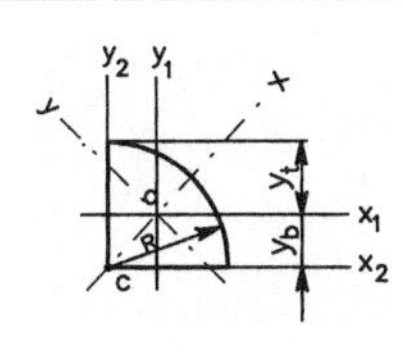

19. Quarter of a CIRCLE

$$A=\frac{\pi R^2}{4}\approx 0.785R^2,\quad y_b=\frac{4R}{3\pi}\approx 0.424R,\quad y_t\approx 0.576R,$$

$$I_x=0.07135R^4,\quad I_y=0.03843R^4,$$

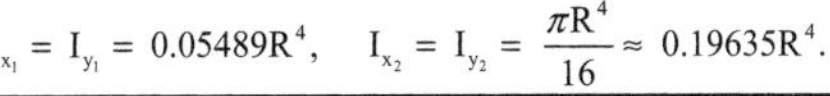

$$I_{x_1}=I_{y_1}=0.05489R^4,\quad I_{x_2}=I_{y_2}=\frac{\pi R^4}{16}\approx 0.19635R^4.$$

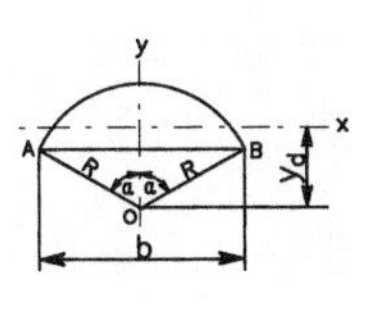

20. Segment of a CIRCLE

$$\hat{\alpha}=\frac{\pi\alpha^0}{180^0},\quad \varphi=2\hat{\alpha}-\sin 2\alpha,\quad k=\frac{4\sin^3\alpha}{3\varphi},\quad b=2R\sin\alpha,\quad s=2R\hat{\alpha},$$

$$A=\frac{R^2\varphi}{2},\quad y_d=kR,\quad I_x=\frac{\varphi R^4}{8}\left(1+3k\cos\alpha\right),\quad I_y=\frac{\varphi R^4}{8}\left(1-k\cos\alpha\right),$$

($\hat{\alpha}$ - in radians measure, α- in degrees).

NOTES

PROPERTIES OF GEOMETRIC SECTIONS

for TENSION, COMPRESSION, and BENDING STRUCTURES

2.5

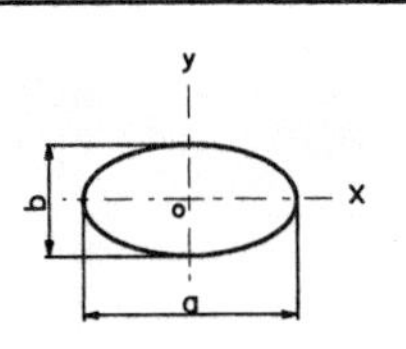

21. ELLIPSE

$$A = \frac{\pi}{4}ab, \quad I_x = \frac{\pi ab^3}{64} = \frac{Ab^2}{16}, \quad I_y = \frac{\pi a^3 b}{64} = \frac{Aa^2}{16},$$

$$S_x = \frac{\pi ab^2}{32} = \frac{Ab}{8}, \quad S_y = \frac{\pi a^2 b}{32} = \frac{Aa}{8},$$

$$r_x = \frac{b}{4}, \quad r_y = \frac{a}{4}$$

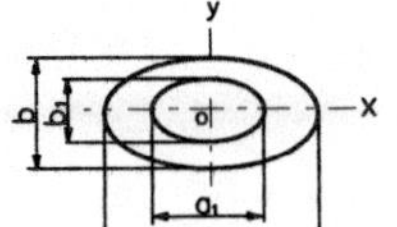

22. HOLLOW ELLIPSE

$$A = \frac{\pi}{4}(ab - a_1 b_1),$$

$$I_x = \frac{\pi}{64}(ab^3 - a_1 b_1^3), \quad I_y = \frac{\pi}{64}(a^3 b - a_1^3 b_1),$$

$$S_x = \frac{\pi}{32b}(ab^3 - a_1 b_1^3), \quad S_y = \frac{\pi}{32a}(a^3 b - a_1^3 b_1)$$

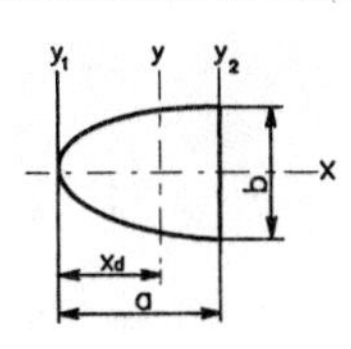

23. Segment of a PARABOLA

$$A = \frac{4ab}{3}, \quad x_d = \frac{3a}{5}, \quad I_x = \frac{4ab^3}{15} = \frac{ab^2}{5},$$

$$I_y = \frac{16a^3 b}{175} = \frac{12Aa^2}{175}, \quad I_{y_1} = \frac{4a^3 b}{7} = \frac{3Aa^2}{7},$$

$$I_{y_2} = \frac{32a^3 b}{105} = \frac{8Aa^2}{35}$$

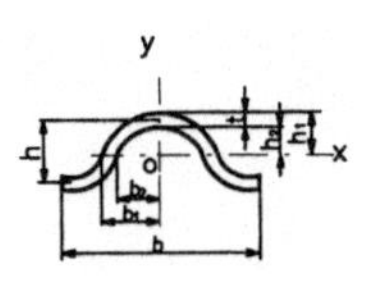

24. STEEL WAVES from parabolic arches

$$A \approx \frac{1}{3}t(2b + 5.2h), \quad b_1 = \frac{1}{4}(b + 2.6t),$$

$$b_2 = \frac{1}{4}(b - 2.6t), \quad h_1 = \frac{1}{2}(h + t),$$

$$h_2 = \frac{1}{2}(h - t), \quad I_x = \frac{64}{105}(b_1 h_1^3 - b_2 h_2^3), \quad S_x \approx \frac{2I_x}{h+t}$$

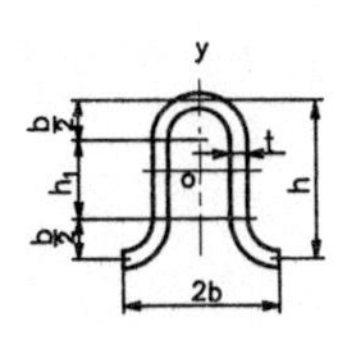

25. STEEL WAVES from circular arches

$$A = (\pi b + 2h)t, \quad h_1 = h - b,$$

$$I_x = \left(\frac{\pi b^3}{8} + b^2 h_1 + \frac{\pi b h_1^2}{4} + \frac{1}{6}h_1^3\right)t,$$

$$S_x = \frac{2I_x}{h+t}$$

NOTES

PROPERTIES OF GEOMETRIC SECTIONS

for TORSION STRUCTURES

2.6

Cross-section	Moment of inertia (I_t)	Elastic section modulus (S_t)	Position of τ_{max} $(\tau_{max} = M_t / S_t)$
d	$I_t = \frac{\pi d^4}{32} = I_p$	$S_t = \frac{\pi d^3}{16}$	At all points of the perimeter
d_1, d_2	$I_t = \frac{\pi}{32} \cdot (d_2^4 - d_1^4) = I_p$	$S_t = \frac{\pi}{16} \cdot \frac{d_2^4 - d_1^4}{d_2}$	At all points of the outside perimeter
d, R	$I_t = 0.1154\ d^4$	$S_t = 0.1888\ d^3$	In the middle of the sides
y, d, o	$I_t = 0.1075\ d^4$	$S_t = 0.1850\ d^3$	In the middle of the sides
a, a	$I_t = 0.1404\ a^4$	$S_t = 0.208\ a^3$	In the middle of the sides
b_2, b_1, b_2, h	$I_t = \frac{h(b_1^4 - b_2^4)}{12(b_1 - b_2)} - 0.21\ b_2^4$	$S_t = \frac{I_t}{b_1}$	In the middle of the long side

W Shape	Angle	Channel	Structural Tee
	$I_t = \eta \cdot \sum_{i=1}^{i=n} \frac{h_i b_i^3}{3}$	$S_t = \frac{I_t}{b_{max}}$	
Ɪ	L	[	⊥
$n = 3, \quad \eta = 1.2$	$n = 2, \quad \eta = 1.0$	$n = 3, \quad \eta = 1.12$	$n = 2, \quad \eta = 1.15$

NOTES

3. BEAMS

Diagrams and Formulas for Various Loading Conditions

NOTES

The formulas provided in Tables 3.1 to 3.10—for determination of support reactions (R), bending moments (M), and shears (V)—are to be used for elastic beams with constant or variable cross-sections.

The formulas for determination of deflection and angles of deflection can only be used for elastic beams with constant cross-sections.

SIMPLE BEAMS

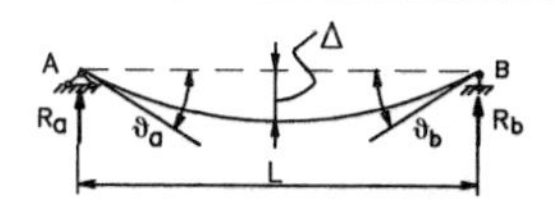

Notes:

$V_1 = R_a, \quad V_2 = R_b$

ϑ_a and ϑ_b in radians

LOADINGS	SUPPORT REACTIONS	BENDING MOMENT	DEFLECTION	ANGLE OF DEFLECTION
L/2, P, L/2; L; Moment; Shear; Mmax; V1; V2	$R_a = \frac{P}{2}$ $R_b = \frac{P}{2}$	$M_{max} = \frac{PL}{4}$ at point of load	$\Delta_{max} = \frac{PL^3}{48EI}$ at point of load	$\vartheta_a = \vartheta_b = \frac{PL^2}{16EI}$
a=ξL, P, b=ξL; L; Moment; Shear; Mmax; V1; V2	$R_a = P\frac{b}{L}$ $R_b = P\frac{a}{L}$	$M_{max} = P\frac{ab}{L}$ at point of load	$\Delta_a = \frac{Pa^2b^2}{3EI \cdot L}$ at point of load	$\vartheta_a = \frac{PL^2}{6EI}(\xi_1 - \xi_1^3)$ $\vartheta_b = \frac{PL^2}{6EI}(\xi - \xi^3)$ $\xi = \frac{a}{L}, \quad \xi_1 = \frac{b}{L}$
a, P, P, a; L; Moment; Shear; Mmax; V1; V2	$R_a = R_b = P$	$M_{max} = Pa$ between loads	$\Delta_{max} = \frac{Pa(3L^2 - 4a^2)}{24EI}$ at center	$\vartheta_a = \vartheta_b = \frac{Pa(L-a)}{2EI}$
L/4, P, L/4, P, L/4, P, L/4; L; Moment; Shear; Mmax; V1; V2	$R_a = \frac{3P}{2}$ $R_b = \frac{3P}{2}$	$M_{max} = \frac{PL}{2}$ at center	$\Delta_{max} = \frac{PL^3}{20.22EI}$ at center	$\vartheta_a = \vartheta_b = 3.75\frac{PL^2}{24EI}$

N O T E S

Table 3.2

Example. Computation of beam

Given. Simple beam $W14\times145$, $L=10$ m

Moment of inertia $I=1710\ \text{in}^4\times2.54^4=71175.6\ \text{cm}^4$

Modulus of elasticity $E=29000\ \text{kip/in}^2=\dfrac{29000\times4.48222}{2.54^2}=20147.6\ \text{kN/cm}^2$

Uniform distribution load $w=5\ \text{kN/m}=0.05\ \text{kN/cm}$

Required. Compute $V=R$, M_{max}, Δ_{max}, $\vartheta=\vartheta_a=\vartheta_b$

Solution.

$$V=R=\frac{wL}{2}=\frac{5\times10}{2}=25\ \text{kN}$$

$$M_{max}=\frac{wL^2}{8}=\frac{5\times10^2}{8}=62.5\ \text{kN}\cdot\text{m}$$

$$\Delta_{max}=\frac{5}{384}\cdot\frac{wL^4}{EI}=\frac{5}{384}\cdot\frac{0.05\times(1000)^4}{20147.6\times71175.6}=0.45\ \text{cm}=4.5\ \text{mm}$$

$$\vartheta=\frac{wL^3}{24EI}=\frac{0.05\times(1000)^3}{24\times20147.6\times71175.6}=1.45\times10^{-3}\,\text{radian}$$

SIMPLE BEAMS

3.2

LOADINGS	SUPPORT REACTIONS	BENDING MOMENT	DEFLECTION	ANGLE OF DEFLECTION
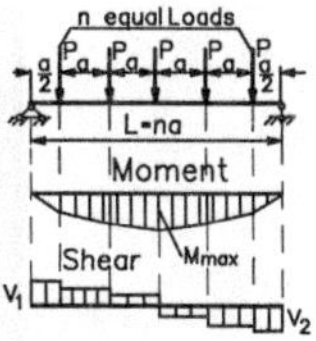 n equal Loads; P; a; a/2; L=na; Moment; Mmax; Shear; V1; V2	$R_a = \frac{Pn}{2}$ $R_b = \frac{Pn}{2}$	n = 4: $M_{max} = \frac{PL}{2}$, $\Delta_{max} = \frac{PL^3}{19.04EI}$	n = 5: $M_{max} = \frac{PL}{1.538}$, $\Delta_{max} = \frac{PL^3}{15.1EI}$ n = 6: $M_{max} = \frac{PL}{1.333}$, $\Delta_{max} = \frac{PL^3}{12.65EI}$	$\vartheta_a = \frac{PL^2}{48EI} \cdot \frac{2n^2+1}{n}$ $\vartheta_b = \frac{PL^2}{48EI} \cdot \frac{2n^2+1}{n}$
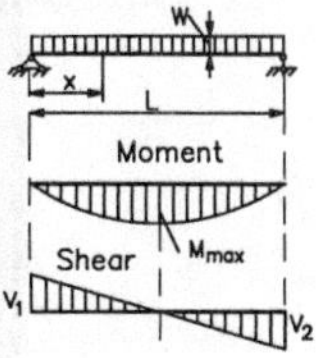 W; x; L; Moment; Mmax; Shear; V1; V2	$R_a = \frac{wL}{2}$ $R_b = \frac{wL}{2}$	$M_{max} = \frac{wL^2}{8}$ at center $M_x = \frac{wx}{2}(L-x)$	$\Delta_{max} = \frac{5}{384} \cdot \frac{wL^4}{EI}$ at center $\Delta_x = \frac{wx\left(L^3 - 2Lx^2 + x^3\right)}{24EI}$	$\vartheta_a = \vartheta_b = \frac{wL^3}{24EI}$
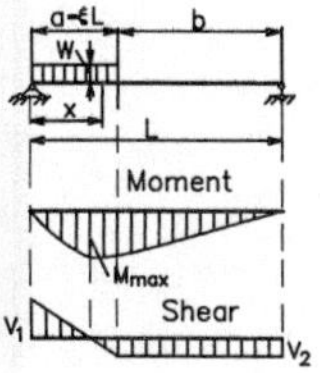 a=ξL; b; W; x; L; Moment; Mmax; Shear; V1; V2	$R_a = \frac{wa}{2}(2-\xi)$ $R_b = \frac{wa}{2} \cdot \xi$ $\xi = \frac{a}{L}$	$M_{max} = \frac{wa^2}{8}(2-\xi)^2$ at $x = \frac{a}{2}(2-\xi)$	$\Delta_a = \frac{wa^3b}{24EI}(4-3\xi)$ at $x = a$	$\vartheta_a = \frac{wa^2L}{6EI}\left(1-\frac{1}{2}\xi\right)^2$ $\vartheta_b = \frac{wa^2L}{12EI}\left(1-\frac{1}{2}\xi^2\right)$
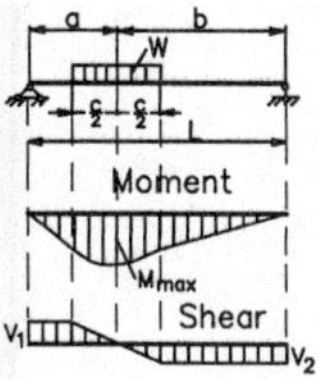 a; b; W; c/2; c/2; L; Moment; Mmax; Shear; V1; V2	$R_a = \frac{wcb}{L}$ $R_b = \frac{wca}{L}$	$M_{max} = \frac{wabc}{L}\left(1-\frac{c}{2L}\right)$ at $x = a + \frac{c(b-a)}{2L}$	$\Delta_a = \left[a\left(2aL - 2a^2 - \frac{c^2}{4}\right) + \frac{c^3L}{64b}\right] \times \frac{R_a}{6EI}$ at $x = a$	$\vartheta_a = \frac{R_a}{24EI} \cdot f_1$ $\vartheta_b = \frac{R_b}{24EI} \cdot f_1$ $f_1 = 4a(L+b) - c^2$

NOTES

SIMPLE BEAMS

<table>
<tr><th>LOADINGS</th><th>SUPPORT REACTIONS</th><th colspan="3">BENDING MOMENT</th><th colspan="3">DEFLECTION</th><th>ANGLE OF DEFLECTION</th></tr>
<tr><td> 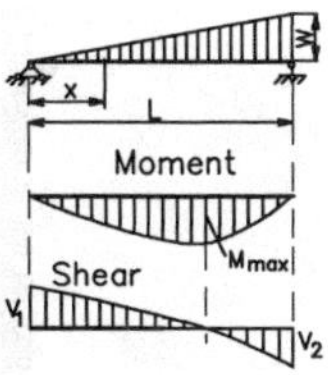w, x, L
Moment
Mmax
Shear
V1, V2</td><td>$R_a = \frac{wL}{6}$
$R_b = \frac{wL}{3}$</td><td colspan="3">$M_{max} = \frac{wL^2}{9\sqrt{3}} = 0.064wL^2$
when $x = 0.577L$</td><td colspan="3">$\Delta_{max} = 0.00652\frac{wL^4}{EI}$
when $x = 0.519L$</td><td>$\vartheta_a = \frac{7}{360} \cdot \frac{wL^3}{EI}$
$\vartheta_b = \frac{8}{360} \cdot \frac{wL^3}{EI}$</td></tr>
<tr><td> 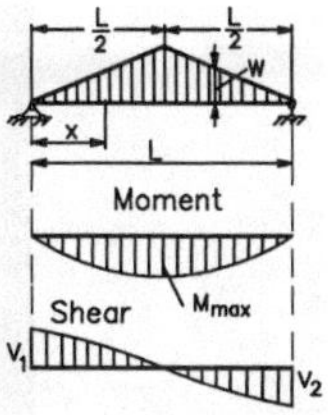L/2, L/2, w, x, L
Moment
Mmax
Shear
V1, V2</td><td>$R_a = R_b = \frac{wL}{4}$</td><td colspan="3">$M_{max} = \frac{wL^2}{12}$
at center</td><td colspan="3">$\Delta_{max} = \frac{wL^2}{120EI}$
at center</td><td>$\vartheta_a = \vartheta_b = \frac{5wL^3}{192EI}$</td></tr>
<tr><td> 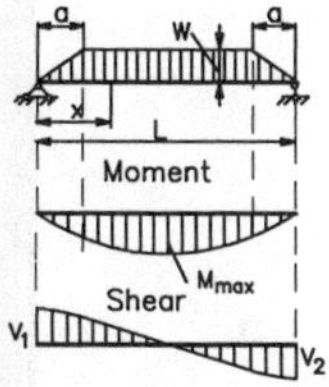a, w, a, x, L
Moment
Mmax
Shear
V1, V2</td><td>$R_a = \frac{w(L-a)}{2}$
$R_b = \frac{w(L-a)}{2}$</td><td colspan="3">$M_{max} = \frac{wL^2}{8} - \frac{wa^2}{6}$
at center</td><td colspan="3">$\Delta_{max} = \frac{5}{384} \cdot \frac{wL^4}{EI} \cdot f_2$
$f_2 = 1 - \frac{8}{5}\xi^2 + \frac{16}{25}\xi^4$
at center</td><td>$\vartheta_a = \vartheta_b = \frac{wL^3}{24EI} \cdot f_3$
$f_3 = 1 - 2\xi^2 + \xi^3$</td></tr>
<tr><td rowspan="4"> 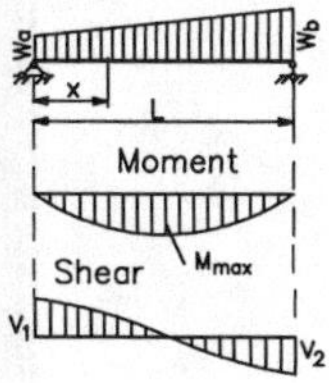wa, wb, x, L
Moment
Mmax
Shear
V1, V2</td><td rowspan="4">$R_a = \frac{2w_a + w_b}{6}L$
$R_b = \frac{w_a + 2w_b}{6}L$</td><td>$\frac{w_a}{w_b} =$</td><td>0.2</td><td>0.4</td><td>0.6</td><td>0.8</td><td>1.0</td><td rowspan="4">$\vartheta_a = \frac{L^3(8w_a + 7w_b)}{360EI}$
$\vartheta_b = \frac{L^3(7w_a + 8w_b)}{360EI}$</td></tr>
<tr><td>$M_{max} =$</td><td>$\frac{w_bL^2}{13.09}$</td><td>$\frac{w_bL^2}{11.30}$</td><td>$\frac{w_bL^2}{9.93}$</td><td>$\frac{w_bL^2}{8.87}$</td><td>$\frac{w_bL^2}{8.00}$</td></tr>
<tr><td>$\frac{x}{L} =$</td><td>0.555</td><td>0.536</td><td>0.520</td><td>0.508</td><td>0.500</td></tr>
<tr><td colspan="3">$\Delta_{max} = (w_a + w_b)L^4$,</td><td colspan="3">when $x = 0.500L$
to $x = 0.519L$</td></tr>
</table>

NOTES

SIMPLE BEAMS and BEAMS OVERHANGING ONE SUPPORT **3.4**

LOADINGS	SUPPORT REACTIONS	BENDING MOMENT	DEFLECTION	ANGLE OF DEFLECTION
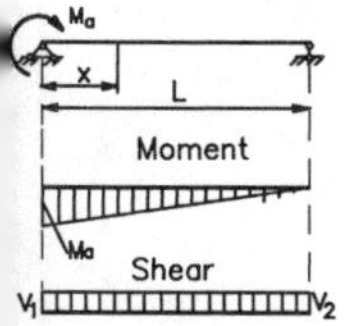 M_a, x, L Moment M_a Shear V_1 V_2	$R_a = \frac{M_a}{L}$ $R_b = -R_a$	$M_{max} = M_a$ when $x = 0$	$\Delta_{max} = \frac{M_a L^2}{15.59EI}$ when $x = 0.423L$ $\Delta = \frac{M_a L^2}{16EI}$ when $x = 0.5L$	$\vartheta_a = \frac{M_a L}{3EI}$ $\vartheta_b = \frac{M_a L}{6EI}$
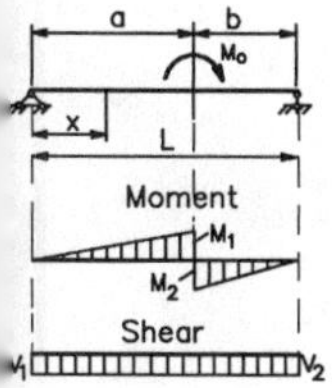 a, b, M_0, x, L Moment M_1, M_2 Shear V_1 V_2	$R_a = -\frac{M_0}{L}$ $R_b = \frac{M_0}{L}$	$M_1 = -M_0 \frac{a}{L}$ $M_2 = M_0 \frac{b}{L}$	$\Delta = \frac{M_0 ab}{3EI}\left(\frac{a-b}{L}\right)$ when $x = a$	$\vartheta_a = -\frac{M_0 L}{6EI} f_4$ $\vartheta_b = \frac{M_0 L}{6EI} f_5$ $f_4 = 1-3\left(\frac{b}{L}\right)^2$ $f_5 = 1-3\left(\frac{a}{L}\right)^2$
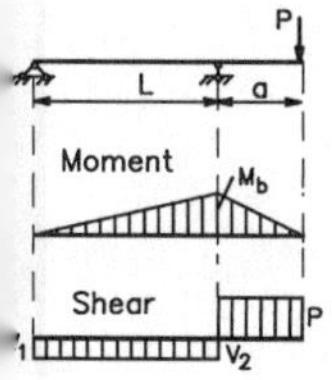 P, L, a Moment M_b Shear P V_1 V_2	$R_a = -P\frac{a}{L}$ $R_b = P\frac{a+L}{L}$	$M_b = -Pa$	For overhang: $\Delta = \frac{Pa^2}{3EI}(L+a)$ Between supports: $\Delta_{max} = -0.0642\frac{PaL^2}{EI}$, $x = 0.577L$	For overhang: $\vartheta = \frac{P(2aL+3a^2)}{6EI}$ $\vartheta_a = -\frac{PaL}{6EI}$ $\vartheta_b = -\frac{PaL}{3EI}$
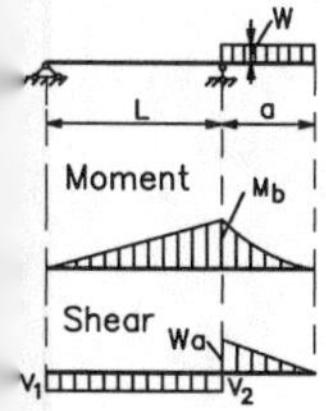 W, L, a Moment M_b Shear Wa V_1 V_2	$R_a = -\frac{wa^2}{2L}$ $R_b = w\left(a+\frac{a^2}{2L}\right)$	$M_b = -\frac{wa^2}{2}$	For overhang: $\Delta = \frac{wa^3}{24EI}(4L+3a)$ Between supports: $\Delta_{max} = -0.0321\frac{wa^2L^2}{EI}$, $x = 0.577L$	For overhang: $\vartheta = \frac{wa^2(a+L)}{6EI}$ $\vartheta_a = -\frac{wa^2L}{12EI}$ $\vartheta_b = -\frac{wa^2L}{6EI}$

NOTES

CANTILEVER BEAMS

3.5

LOADINGS	REACTION (at fixed end)	BENDING MOMENT (at fixed end)	DEFLECTION (at free end)	ANGLE OF DEFLECTION (at free end)
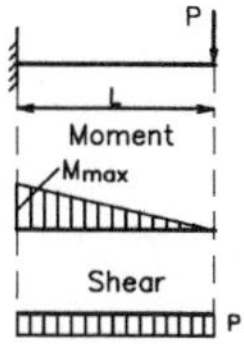 P; L; Moment; Mmax; Shear; P	$R = P$	$M_{max} = -PL$	$\Delta_{max} = \frac{PL^3}{3EI}$	$\vartheta = \frac{PL^2}{2EI}$
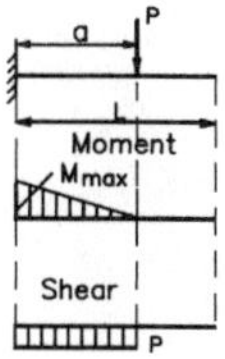 a; P; L; Moment; Mmax; Shear; P	$R = P$	$M_{max} = -Pa$	$\Delta_{max} = \frac{Pa^2}{6EI}(3L - a)$	$\vartheta = \frac{Pa^2}{2EI}$
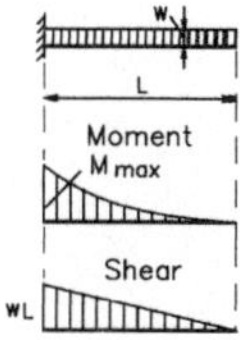 w; L; Moment; M max; Shear; wL	$R = wL$	$M_{max} = -\frac{wL^2}{2}$	$\Delta_{max} = \frac{wL^4}{EI}$	$\vartheta = \frac{wL^3}{6EI}$
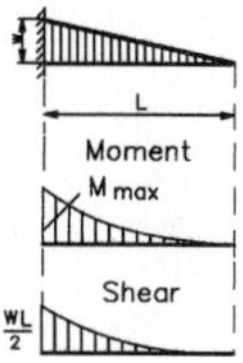 w; L; Moment; M max; Shear; wL/2	$R = \frac{wL}{2}$	$M_{max} = -\frac{wL^2}{6}$	$\Delta_{max} = \frac{wL^4}{30EI}$	$\vartheta = \frac{wL^3}{24EI}$

NOTES

BEAMS FIXED AT ONE END, SUPPORTED AT OTHER

3.6

LOADINGS	SUPPORT REACTIONS	BENDING MOMENTS AND DEFLECTION
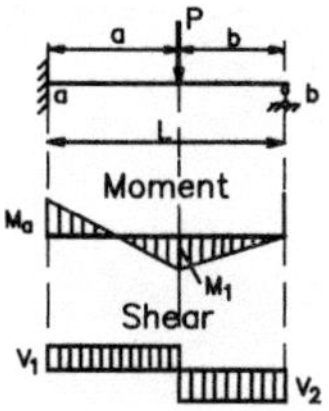 a, P, b, L, Moment, M_a, M_1, Shear, V_1, V_2	$R_a = \frac{Pb}{2L^3}\left(3L^2 - b^2\right)$ $R_b = \frac{Pa^2}{2L^3}\left(b + 2L\right)$	$M_a = -\frac{Pab}{2L^2}\left(L + b\right)$, at fixed end $M_1 = R_b b$, at point of load $\Delta_1 = \frac{Pa^2b^2\left(3a + 4b\right)}{12L^3EI}$, at point of load
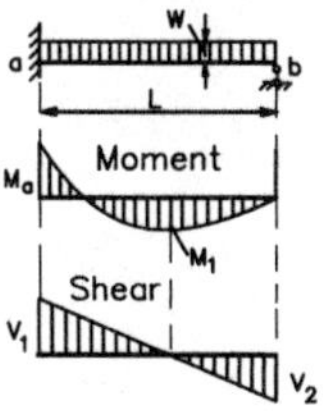 a, w, b, L, Moment, M_a, M_1, Shear, V_1, V_2	$R_a = \frac{5}{8}wL$ $R_b = \frac{3}{8}wL$	$M_a = -\frac{wL^2}{8}$, at fixed end $M_1 = \frac{9}{128}wL^2$, at $x = 0.625L$ $\Delta_{max} = \frac{wL^4}{185EI}$, at $x = 0.579L$ $\Delta = \frac{wL^4}{192EI}$, at $x = \frac{L}{2}$
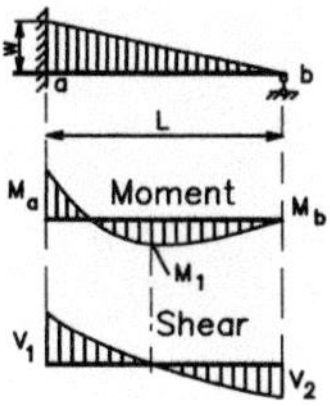 w, a, b, L, Moment, M_a, M_b, M_1, Shear, V_1, V_2	$R_a = \frac{2}{5}wL$ $R_b = \frac{1}{10}wL$	$M_a = -\frac{wL^2}{15}$, at fixed end $M_1 = \frac{wL^2}{33.6}$, at $x = 0.553L$ $\Delta_{max} = \frac{wL^4}{419EI}$, at $x = 0.553L$ $\Delta = \frac{wL^4}{426.6EI}$, at $x = \frac{L}{2}$
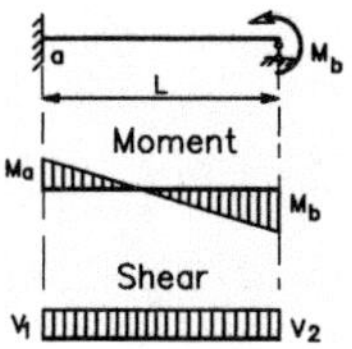 a, M_b, L, Moment, M_a, M_b, Shear, V_1, V_2	$R_a = \frac{3}{2}\cdot\frac{M_b}{L}$ $R_b = -\frac{3}{2}\cdot\frac{M_b}{L}$	$M_a = -\frac{M_b}{2}$, at fixed end $\Delta_{max} = \frac{M_bL^2}{27EI}$, at $x = \frac{2}{3}L$

NOTES

BEAMS FIXED AT ONE END, SUPPORTED AT OTHER

3.7

LOADINGS	SUPPORT REACTIONS	BENDING MOMENT (AT FIXED END)
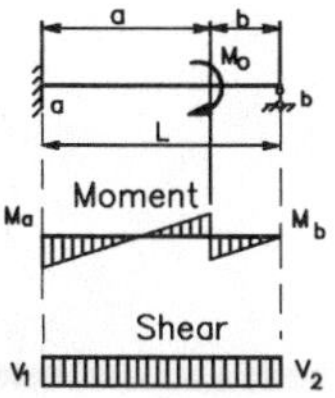 a, b, M_0, a, b, L; Moment: M_a, M_b; Shear: V_1, V_2	$R_a = -\frac{3M_0(L^2-b^2)}{2L^3}$ $R_b = \frac{3M_0(L^2-b^2)}{2L^3}$	$M_a = \frac{M_0}{2}\left[1-3\left(\frac{b}{L}\right)^2\right]$, when $b < 0.577L$ $M_a = 0$, when $b = 0.577L$ $M_a = -\frac{M_0}{2}\left[1-3\left(\frac{b}{L}\right)^2\right]$, when $b > 0.577L$
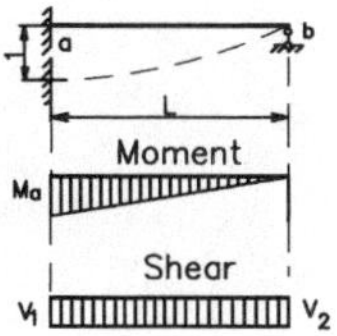 1, a, b, L; Moment: M_a; Shear: V_1, V_2	$R_a = -\frac{3EI}{L^3}$ $R_b = \frac{3EI}{L^3}$	$M_a = \frac{3EI}{L^2}$
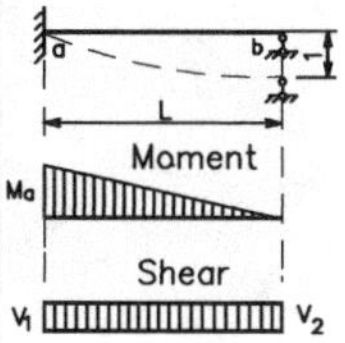 a, b, 1, L; Moment: M_a; Shear: V_1, V_2	$R_a = \frac{3EI}{L^3}$ $R_b = -\frac{3EI}{L^3}$	$M_a = -\frac{3EI}{L^2}$
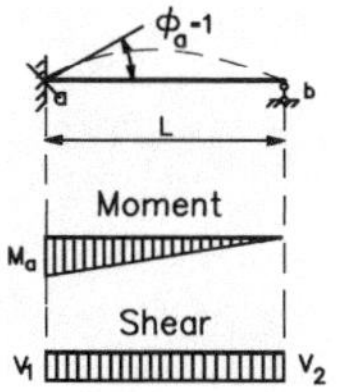 $\phi_a = 1$, a, b, L; Moment: M_a; Shear: V_1, V_2	$R_a = \frac{3EI}{L^2}$ $R_b = -\frac{3EI}{L^2}$	$M_a = -\frac{3EI}{L}$

NOTES

BEAMS FIXED AT BOTH ENDS

3.8

LOADINGS	SUPPORT REACTIONS	BENDING MOMENTS AND DEFLECTION
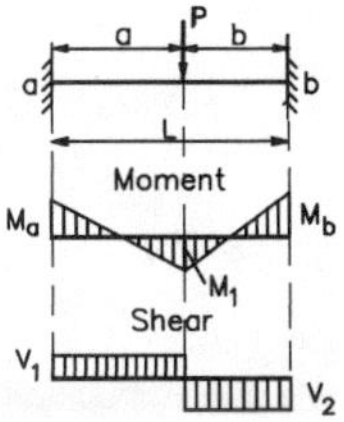 a, P, b, L, Moment, M_a, M_b, M_1, Shear, V_1, V_2	$R_a = \frac{P(3a+b)b^2}{L^3}$ $R_b = \frac{P(a+3b)a^2}{L^3}$	$M_a = -\frac{Pab^2}{L^2}$, $M_b = -\frac{Pa^2b}{L^2}$ $M_1 = \frac{2Pa^2b^2}{L^3}$, at point of load $\Delta_1 = \frac{Pa^3b^3}{3L^3EI}$, at point of load
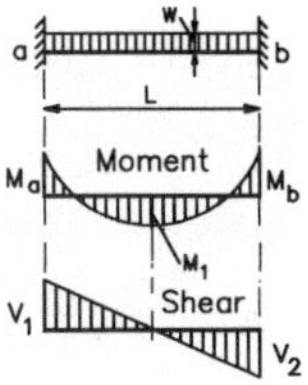 a, w, b, L, Moment, M_a, M_b, M_1, Shear, V_1, V_2	$R_a = R_b = \frac{wL}{2}$	$M_a = M_b = -\frac{wL^2}{12}$ $M_1 = \frac{wL^2}{24}$, at center $\Delta_{max} = \frac{wL^4}{384EI}$, at center
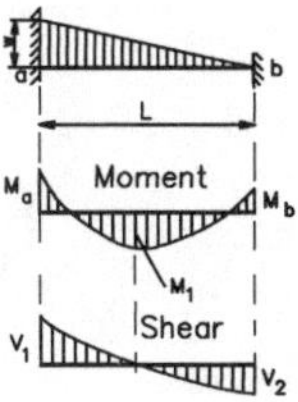 w, a, b, L, Moment, M_a, M_b, M_1, Shear, V_1, V_2	$R_a = \frac{7}{20}wL$ $R_b = \frac{3}{20}wL$	$M_a = -\frac{wL^2}{20}$, $M_b = -\frac{wL^2}{30}$ $M_1 = \frac{wL^2}{46.6}$, at $x = 0.452L$ $\Delta_{max} = \frac{wL^4}{764EI}$, at $x = 0.475L$ $\Delta = \frac{wL^4}{768EI}$, at $x = \frac{L}{2}$
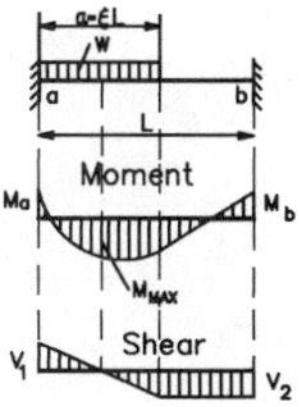 a=ξL, w, a, b, L, Moment, M_a, M_b, M_{MAX}, Shear, V_1, V_2	$R_a = \frac{wa(L-0.5a)}{L} - \frac{M_a - M_b}{L}$ $R_b = \frac{wa^2}{2L} + \frac{M_a - M_b}{L}$	$M_a = -\frac{wa^2}{6}(3 - 4\xi + 1.5\xi^2)$ $M_b = -\frac{wa^2}{3}(\xi - 0.75\xi^2)$ $\xi = \frac{a}{L}$

NOTES

BEAMS FIXED AT BOTH ENDS

3.9

LOADINGS	SUPPORT REACTIONS	BENDING MOMENTS (AT FIXED ENDS)
a, c=ξL, a, a, b, L, Moment, M_a, M_b, M_1, Shear, V_1, V_2	$R_a = R_b = \frac{wc}{2}$	$M_a = M_b = -\frac{wcL}{24}\left(3-\xi^2\right)$ $\xi = \frac{c}{L}$ $M_1 = \frac{wcL}{4}\left(1-\frac{1}{2}\xi\right) - \frac{wcL}{24}\left(3-\xi^2\right)$ at center
a, b, a, b, M_o, L, Moment, M_a, M_b, Shear, V_1, V_2	$R_a = -\frac{6M_0ab}{L^3}$ $R_b = \frac{6M_0ab}{L^3}$	$M_a = \frac{M_0 b}{L^2}\left(2a-b\right)$ $M_b = \frac{M_0 b}{L^2}\left(a-2b\right)$ When $x = \frac{L}{3}$: $M_a = 0$, $M_b = -\frac{M_0}{3}$
a, b, 1, Moment, M_a, M_b, Shear, V_1, V_2	$R_a = \frac{12EI}{L^3}$ $R_b = -\frac{12EI}{L^3}$	$M_a = -\frac{6EI}{L^2}$ $M_b = \frac{6EI}{L^2}$
φ, ϕ_a=1, a, b, Moment, M_a, M_b, Shear, V_1, V_2	$R_a = \frac{6EI}{L^2}$ $R_b = -\frac{6EI}{L^2}$	$M_a = -\frac{4EI}{L}$ $M_b = \frac{2EI}{L}$

NOTES

Support Reaction (R), Shear (V), Bending Moment (M), Deflection (Δ)

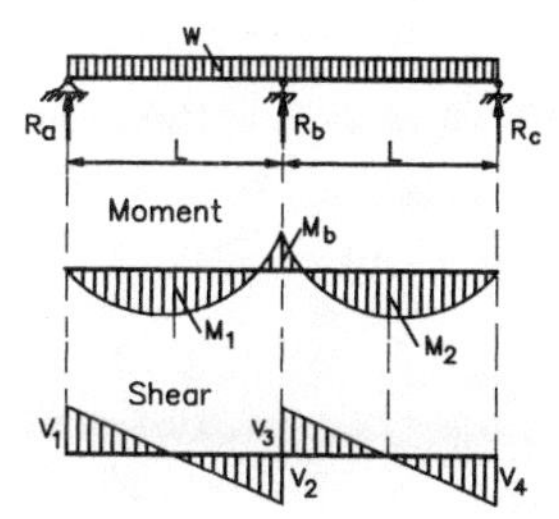

$$R_a = V_1 = 0.375wL$$

$$R_b = V_2 + V_3 = 1.250wL, \quad V_2 = V_3 = 0.625wL$$

$$R_c = V_4 = 0.375wL$$

$$M_1 = M_2 = 0.070wL^2,$$
at 0.375L from R_a and R_c

$$M_b = -\ 0.125wL^2$$

$$\Delta = 0.0052\frac{wL^4}{EI},$$ in the middle of the spans

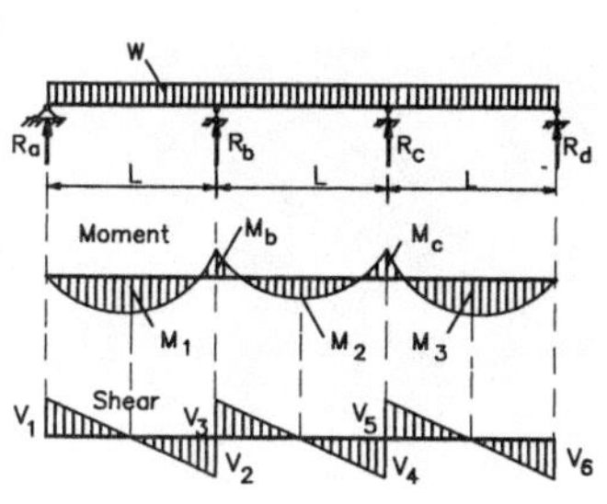

$$R_a = V_1 = 0.400wL, \quad R_d = V_6 = 0.400wL$$

$$R_b = R_c = 1.100wL, \ V_2 + V_5 = 0.600wL,$$
$$V_3 = V_4 = 0.500wL$$

$$M_1 = M_3 = 0.080wL^2,$$ at 0.400L from R_a and R_d

$$M_2 = 0.025wL^2, \quad M_b = M_c = -\ 0.100wL^2$$

$$\Delta_{max} = 0.0069\frac{wL^4}{EI},$$ at 0.446L from R_a and R_d

$$\Delta = 0.00675\frac{wL^4}{EI},$$ in the middle of spans 1 and 3

$$\Delta = 0.00052\frac{wL^4}{EI},$$ in the middle of span 2

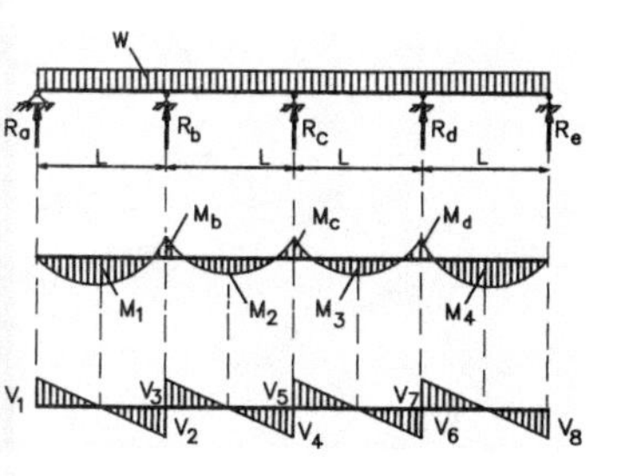

$$R_a = R_e = 0.393wL, \quad R_b = R_d = 1.143wL,$$
$$R_c = 0.928wL$$

$$V_1 = V_8 = 0.393wL, \quad V_2 = V_7 = 0.607wL,$$
$$V_3 = V_6 = 0.536wL, \quad V_4 = V_5 = 0.464wL.$$

$M_1 = M_4 = 0.0772wL^2$, at 0.393L from R_a and R_e

$M_2 = M_3 = 0.0364wL^2$, at 0.536L from R_b and R_d

$M_b = M_d = -\ 0.1071wL^2$, $\quad M_c = -\ 0.0714wL^2$

$M_1 = M_4 = 0.0772wL^2$, at 0.393L from R_a and R_e

$$\Delta_{max} = 0.0065\frac{wL^4}{EI},$$ at 0.440L from R_a and R_e

N O T E S

Table 3.11 is provided for computing bending moments at the supports of elastic continuous beams with equal spans and flexural rigidity along the entire length.

The bending moments resulting from settlement of supports are summated with the bending moments due to acting loads.

Table 3.11 Continuous beams

Example. Settlement of beam support

Given. Three equal spans continuous beam $W12\times35, \quad L = 6.0 \text{ m}$

Moment of inertia $I_z = 285 \text{ in}^4 \times 2.54^4 = 11862.6 \text{ cm}^4$

Modulus of elasticity $E = 29000 \text{ kip/in}^2 = \dfrac{29000\times4.48222}{2.54^2} = 20147.6 \text{ kN/cm}^2$

Settlement of support B: $\Delta_B = 0.8 \text{ cm}$

Required. Compute bending moments M_B and M_C

Solution.

$$M_B = k_B \frac{EI_z}{L^2}\cdot\Delta_B = 3.6\frac{20147.6\times11862.6}{(600)^2}\times0.8 = 1912.0 \text{ kN}\cdot\text{cm} = 19.12 \text{ kN}\cdot\text{m}$$

$$M_C = k_C \frac{EI_z}{L^2}\cdot\Delta_B = -2.4\frac{20147.6\times11862.6}{(600)^2}\times0.8 = -1274.7 \text{ kN}\cdot\text{cm} = -12.75 \text{ kN}\cdot\text{m}$$

CONTINUOUS BEAMS

SETTLEMENT OF SUPPORT

3.11

Bending moment at support :

$$M = k\frac{EI_z}{L^2}\cdot\Delta\,,\quad \text{where } k = \text{coefficient}\,,$$

$$\Delta = \text{settlement of support.}$$

CONTINUOUS BEAM	Bending moment	SUPPORT A	B	C	D	E	F
		COEFFICIENT K					
TWO EQUAL SPANS (A, B, C; 2 x L)	M_B =	−1.500	3.000	−1.500			
THREE EQUAL SPANS (A, B, C, D; 3 x L)	M_B =	−1.600	3.600	−2.400	0.400		
	M_C =	0.400	−2.400	3.600	−1.600		
FOUR EQUAL SPANS (A, B, C, D, E; 4 x L)	M_B =	−1.607	3.643	−2.571	0.643	−0.107	
	M_C =	0.429	−2.571	4.286	−2.571	0.429	
	M_D =	−0.107	0.643	−2.571	3.643	−1.607	
FIVE EQUAL SPANS (A, B, C, D, E, F; 5 x L)	M_B =	−1.608	3.645	−2.583	0.688	−0.172	0.029
	M_C =	0.431	−2.584	4.335	−2.756	0.689	−0.115
	M_D =	−0.115	0.689	−2.756	4.335	−2.584	0.431
	M_E =	0.029	−0.172	0.688	−2.583	3.645	−1.608

N O T E S

Table 3.12

Example. Moving concentrated loads

Given. Simple beam, $L = 30$ m

$P_1 = 40$ kN, $P_2 = 80$ kN, $P_3 = 120$ kN, $P_4 = 100$ kN, $P_5 = 80$ kN, $\sum P_i = 420$ kN

$a = 4$ m, $b = 3$ m, $c = 3$ m, $d = 2$ m

Required. Compute maximum bending moment and maximum end shear

Solution. Center of gravity of loads (off load P_1):

Bending moment

$$\sum(P_i \cdot x_i)/\sum P_i = (80\times4+120\times7+100\times10+80\times14)/420 = 3280/420 = 7.8 \text{ m}$$

$$e = 7.8-(3+4) = 0.8 \text{ m}, \quad e/2 = 0.4 \text{ m}$$

$$R_A = \sum P_i \times\left(\frac{L}{2}-\frac{e}{2}\right)/L = 420(15-0.4)/30 = 204.4 \text{ kN}$$

$$M_{max} = R_A \cdot\left(\frac{L}{2}-\frac{e}{2}\right)-\left[P_1(a+b)+P_2 b\right] = 204.4\times(15-0.4)-\left[40\times(4+3)+80\times3\right] = 2464.2 \text{ kN}\cdot\text{m}$$

End shear

Load P_1 passes off the span and P_2 moves over the left support

$$\Delta V_1 = \frac{\sum P_i \cdot a}{L} - P_1 = \frac{420\times4}{30} - 40 = +16 > 0$$

Load P_2 passes off the span and P_3 moves over the left support

$$\Delta V_2 = \frac{\sum P_i \cdot b}{L} - P_2 = \frac{420\times3}{30} - 80 = -38 < 0$$

For maximum end shear load P_2 is placed over the left support

$$V_{max} = P_2 + \left[P_3(L-b)+P_4(L-b-c)+P_5(L-b-c-d)\right]/L$$

$$= 80+\left[120\times(30-3)+100(30-3-3)+80(30-3-3-2)\right]/30$$

$$= 80+7240/30 = 326.7 \text{ kN}$$

Maximum bending moment

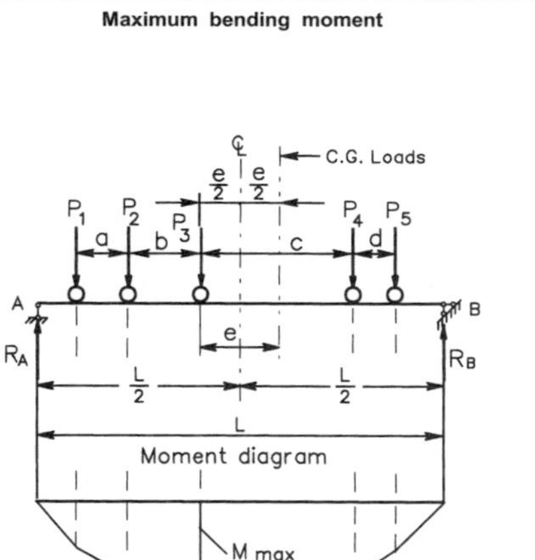

Maximum bending moment caused in a beam by a series of moving concentrated loads occurs when the center of gravity (C.G.) of all the loads and the load nearest to it (P_3 in this example) are on opposite sides of, and the same distance $\left(\frac{e}{2}\right)$ from, the center of the beam.

Maximum end shear

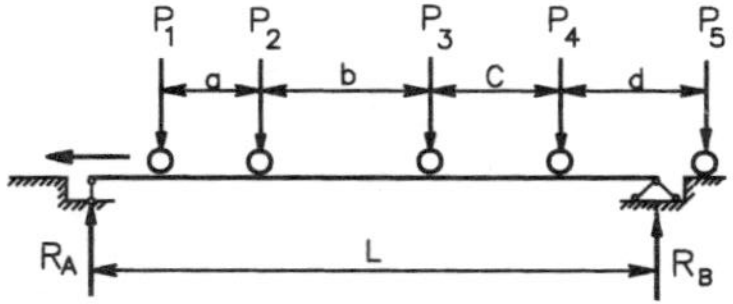

Maximum end shear in a simple beam equals the reaction when one of the moving concentrated loads is at the support.

Moving loads are sequentially placed over the support, and the following expressions are evaluated:

$$\Delta V_1 = \frac{\sum P \cdot a}{L} - P_1 \,, \quad \Delta V_2 = \frac{\sum P \cdot b}{L} - P_2 \; \ldots \,,$$

where: $\sum P$ is the sum of the loads remaining on the beam at any time.

If $\Delta V > 0$, the shear has increased.

If $\Delta V < 0$, the shear has decreased.

Maximum end shear occurs when the first load to produce $\Delta V < 0$ is placed over the support.

NOTES

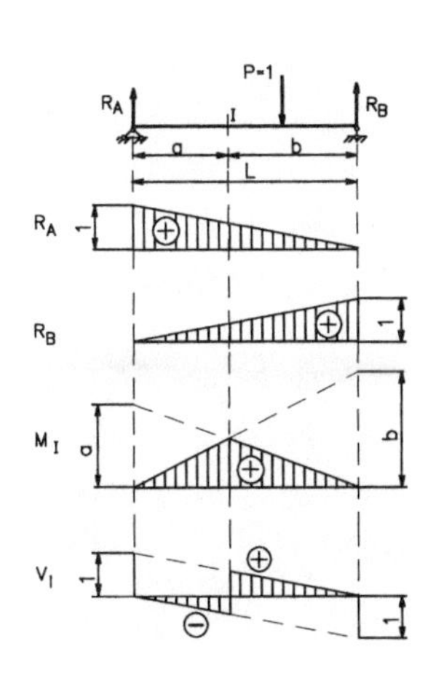

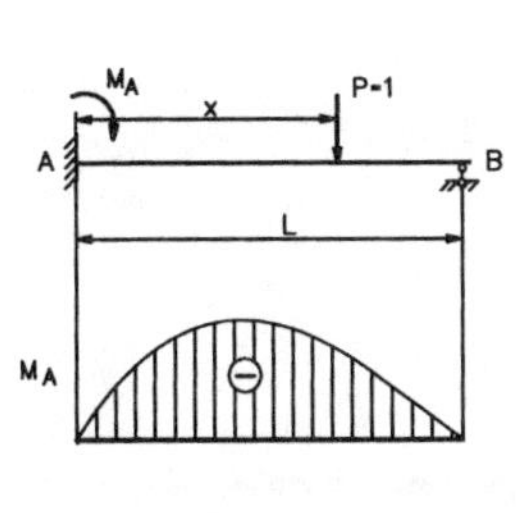

$$M_A = \alpha_x \times L \times P$$

x/L	0.1	0.2	0.3	0.4	0.5
α_x	0.086	0.144	0.178	0.192	0.188
x/L	0.6	0.7	0.8	0.9	1.0
α_x	0.168	0.136	0.096	0.050	0.0

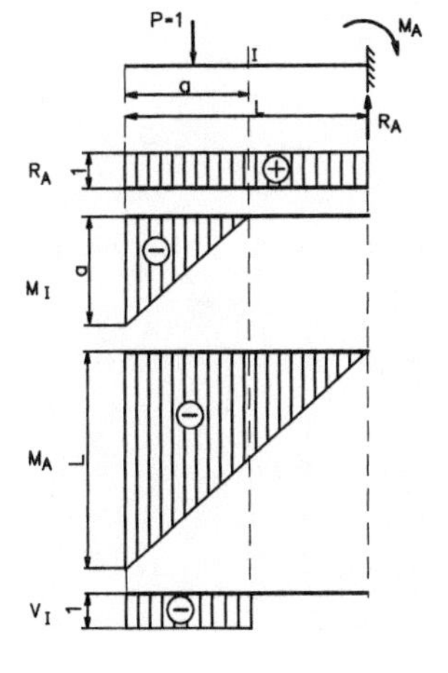

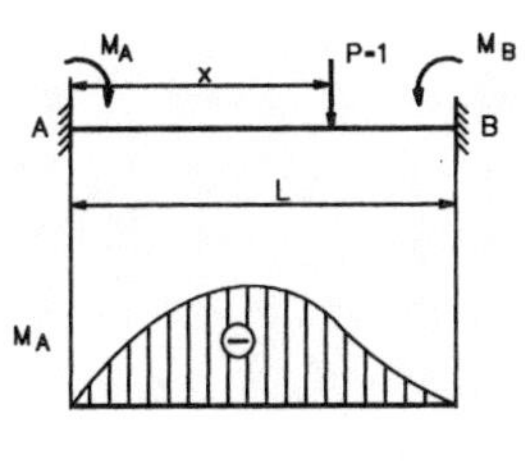

$$M_A = \alpha_x \times L \times P$$

x/L	0.1	0.2	0.3	0.4	0.5
α_x	0.081	0.128	0.147	0.144	0.125
x/L	0.6	0.7	0.8	0.9	1.0
α_x	0.096	0.063	0.032	0.009	0.0

NOTES

BEAMS

INFLUENCE LINES (EXAMPLES)

3.14

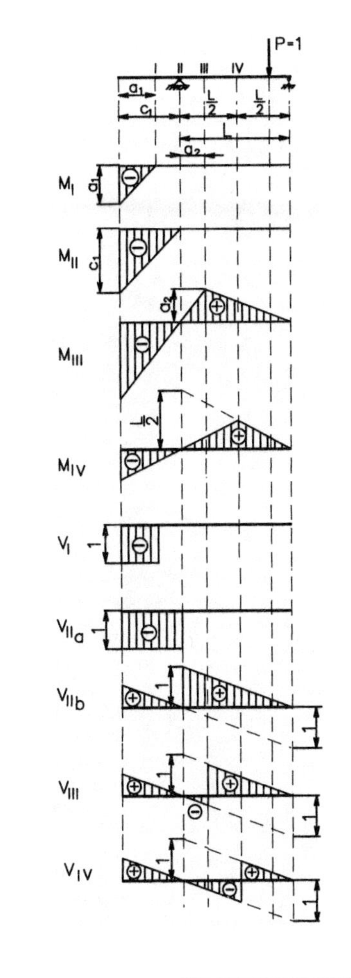

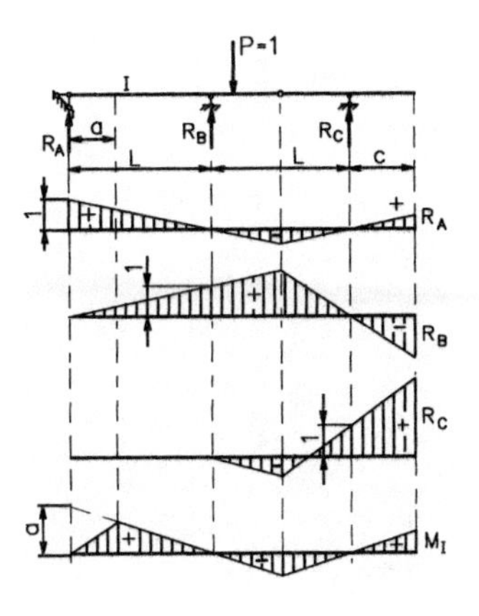

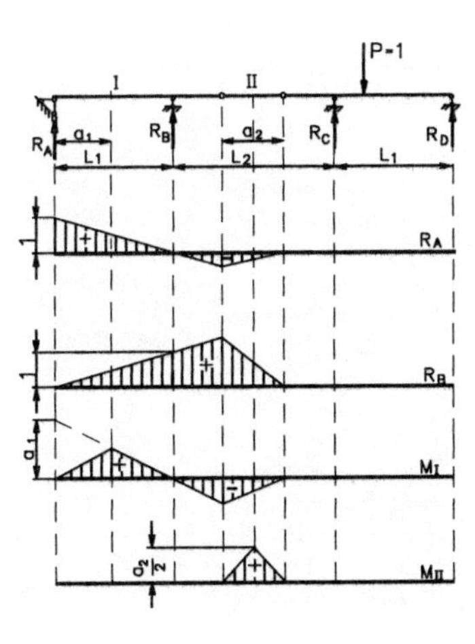

NOTES

(EXAMPLES)

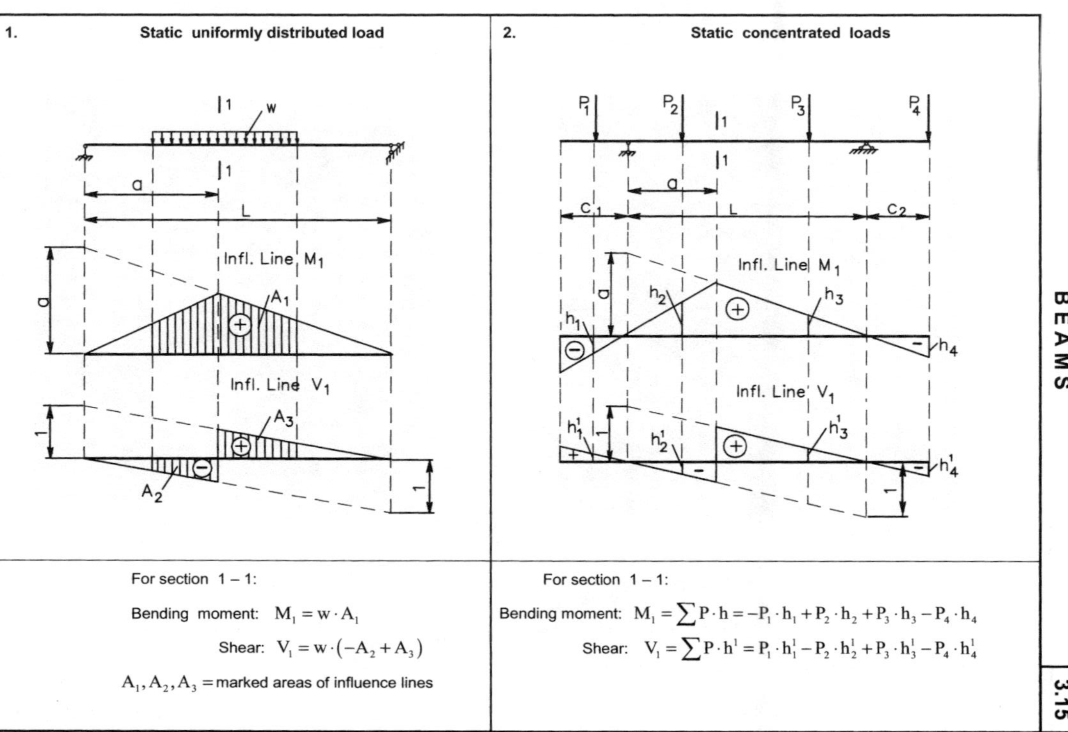

1. Static uniformly distributed load

For section 1 – 1:

Bending moment: $M_1 = w \cdot A_1$

Shear: $V_1 = w \cdot (-A_2 + A_3)$

A_1, A_2, A_3 = marked areas of influence lines

2. Static concentrated loads

For section 1 – 1:

Bending moment: $M_1 = \sum P \cdot h = -P_1 \cdot h_1 + P_2 \cdot h_2 + P_3 \cdot h_3 - P_4 \cdot h_4$

Shear: $V_1 = \sum P \cdot h^1 = P_1 \cdot h_1^1 - P_2 \cdot h_2^1 + P_3 \cdot h_3^1 - P_4 \cdot h_4^1$

NOTES

COMPUTATION OF BENDING MOMENT AND SHEAR USING INFLUENCE LINES (EXAMPLES)

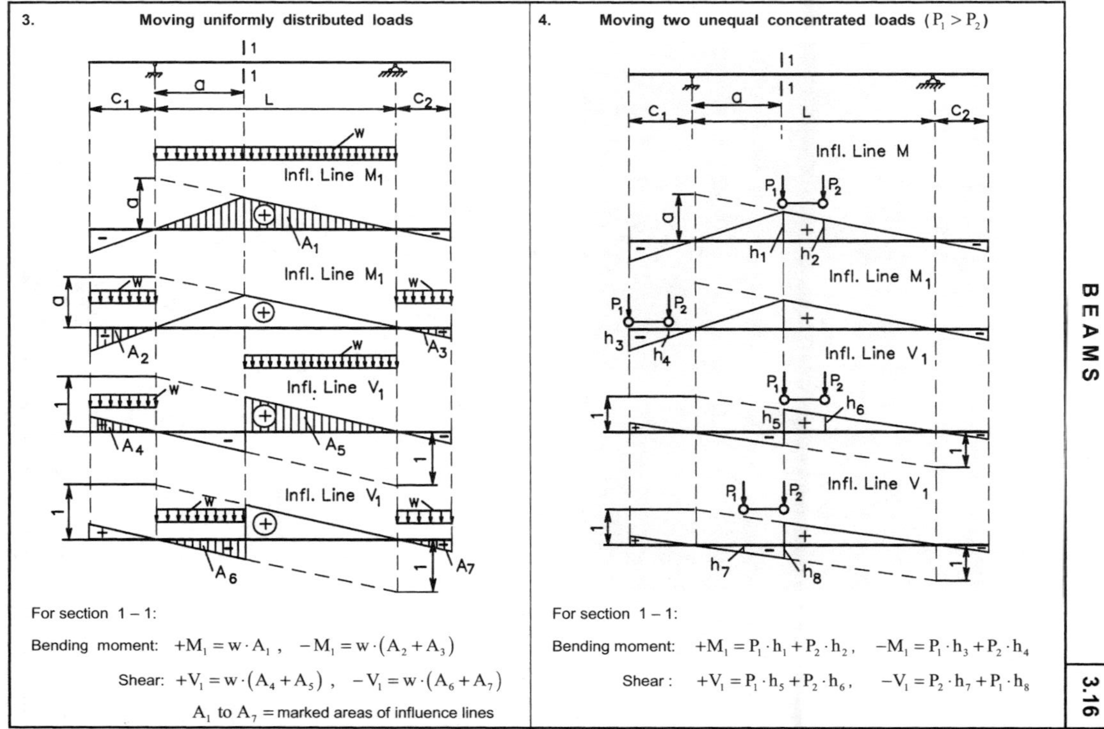

3. For section 1 – 1:

Bending moment: $+M_1 = w \cdot A_1$, $-M_1 = w \cdot (A_2 + A_3)$

Shear: $+V_1 = w \cdot (A_4 + A_5)$, $-V_1 = w \cdot (A_6 + A_7)$

A_1 to A_7 = marked areas of influence lines

4. For section 1 – 1:

Bending moment: $+M_1 = P_1 \cdot h_1 + P_2 \cdot h_2$, $-M_1 = P_1 \cdot h_3 + P_2 \cdot h_4$

Shear: $+V_1 = P_1 \cdot h_5 + P_2 \cdot h_6$, $-V_1 = P_2 \cdot h_7 + P_1 \cdot h_8$

NOTES

4. FRAMES

Diagrams and Formulas for Various Static Loading Conditions

NOTES

The formulas presented in Tables 4.1–4.5 are used for analysis of elastic frames and allow computation of bending moments at corner sections of frame girders and posts. Bending moments at other sections of frame girders and posts can be computed using the formulas provided below.

For girders:

$$\text{If} \quad M_c > M_d, \quad M_{g(x)} = M^0_{g(x)} - \left[\frac{M_c - M_d}{L}(L - x) + M_d\right]$$

$$\text{If} \quad M_c < M_d, \quad M_{g(x)} = M^0_{g(x)} - \left[\frac{M_d - M_c}{L}x + M_c\right]$$

$$\text{If} \quad M_c = M_d = M_s, \quad M_{g(x)} = M^0_{g(x)} - M_s$$

For posts:

$$M_{p(x)} = M^0_{p(x)} - \left(H \cdot x - M_{a(b)}\right)$$

Where: $M^0_{g(x)}$ and $M^0_{p(x)}$ represent, respectively, for frame girders and posts the bending moments in the corresponding simple beam due to the acting load.

x is the distance from the section under consideration to corner c (for the girder) and support a or b (for a post).

DIAGRAMS and FORMULAS for VARIOUS STATIC LOADING CONDITIONS

FRAMES

4.1

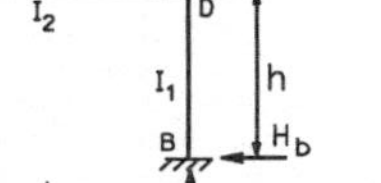
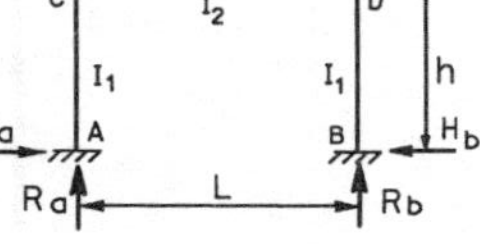
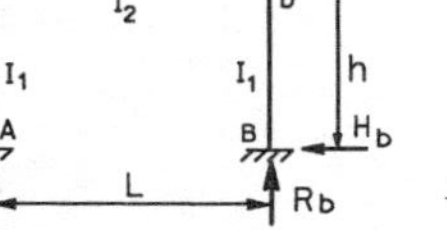
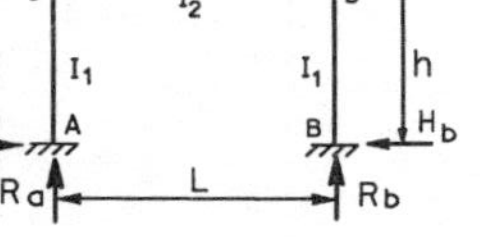

$$k = \frac{I_2 h}{I_1 L}$$

+M = Tension on inside of frame

1

$$H = \frac{wL^2}{4h(k+2)}$$

$$M_a = M_b = \frac{wL^2}{12(k+2)}$$

$$M_c = M_d = -Hh + \frac{wL^2}{12(k+2)}$$

2

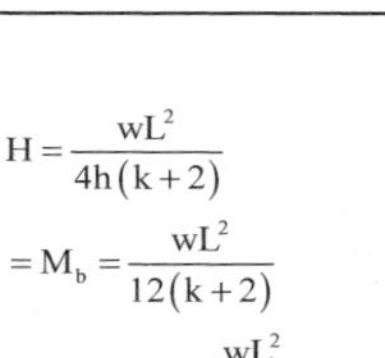

$$H_a = wh - H_b$$

$$H_b = \frac{wh}{8} \cdot \frac{2k+3}{k+2}$$

$$R_b = -R_a = \frac{wh^2}{L} \cdot \frac{k}{6k+1}$$

$$M_a = -\frac{wh^2}{24}\left(\frac{7k+15}{k+2} - \frac{12k}{6k+1}\right)$$

$$M_b = -M_a$$

$$M_c = H_a h - 0.5wh^2 - M_a$$

$$M_d = -H_b h + M_b$$

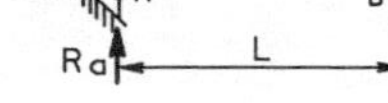
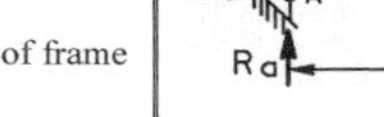

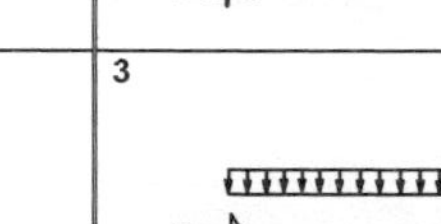
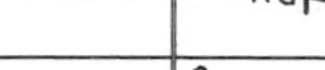

$$k = \frac{I_2 h}{I_1 L}$$

+M = Tension on inside of frame

3

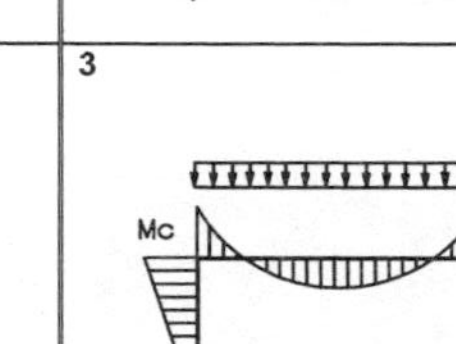

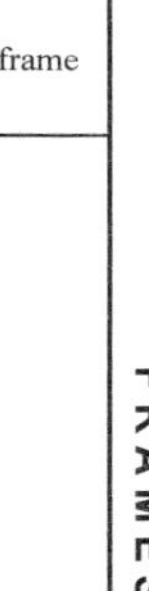
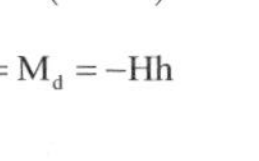

$$H = \frac{wL^2}{4h(2k+3)}$$

$$M_c = M_d = -Hh$$

4

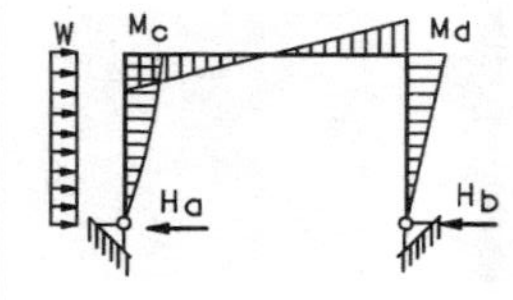

$$H_a = wh - H_b$$

$$H_b = \frac{wh}{8} \cdot \frac{6+5k}{2k+3}$$

$$M_c = H_a h - 0.5wh^2$$

$$M_d = -H_b h$$

NOTES

Example. Analysis of frame

Given. Frame 5 in Table 4.5, $L = 12\text{ m}, \quad h = 3\text{ m}$

Posts $W10\times45, \; I_1 = 248\text{ in}^4 \times 2.54^4 = 10322\text{ cm}^4$

Girder $W14\times82, \; I_2 = 882\text{ in}^4 \times 2.54^4 = 36712\text{ cm}^4$

Load $P = 20\text{ kN}, \quad a = 4\text{ m}, \quad b = 8\text{ m}$

Required. Compute support reactions and bending moments

Solution. $k = \dfrac{I_2 h}{I_1 L} = \dfrac{36712\times3}{10322\times12} = 0.889, \quad \xi = \dfrac{a}{L} = \dfrac{4}{12} = 0.333$

$$H = \frac{3}{2}\cdot\frac{Pab}{hL(k+2)} = \frac{3}{2}\cdot\frac{20\times4\times8}{3\times12(0.889+2)} = 9.23\text{ kN}$$

$$R_a = \frac{Pb}{L}\cdot\frac{1+\xi-2\xi^2+6k}{6k+1} = 13.57\text{ kN}$$

$$R_b = P - R_a = 20 - 13.57 = 6.43\text{ kN}$$

$$M_a = \frac{Pab}{2L}\cdot\frac{5k-1+2\xi(k+2)}{(k+2)(6k+1)} = 7.813\text{ kN}\cdot\text{m}$$

$$M_b = R_a L + M_a - Pb = 13.57\times12 + 7.813 - 20\times8 = 10.653\text{ kN}\cdot\text{m}$$

$$M_c = -Hh + M_a = -9.23\times3 + 7.813 = -19.877\text{ kN}\cdot\text{m}$$

$$M_d = -Hh + M_b = -9.23\times3 + 10.653 = -17.037\text{ kN}\cdot\text{m}$$

Bending moment at point of load

$$M_g = M_g^0 - \left[\frac{M_c - M_d}{L}(L-a) + M_d\right], \quad M_g^0 = \frac{Pab}{L}$$

$$M_g = \frac{20\times4\times8}{12} - \left[\frac{19.877-17.037}{12}(12-4) + 17.037\right] = 34.403\text{ kN}\cdot\text{m}$$

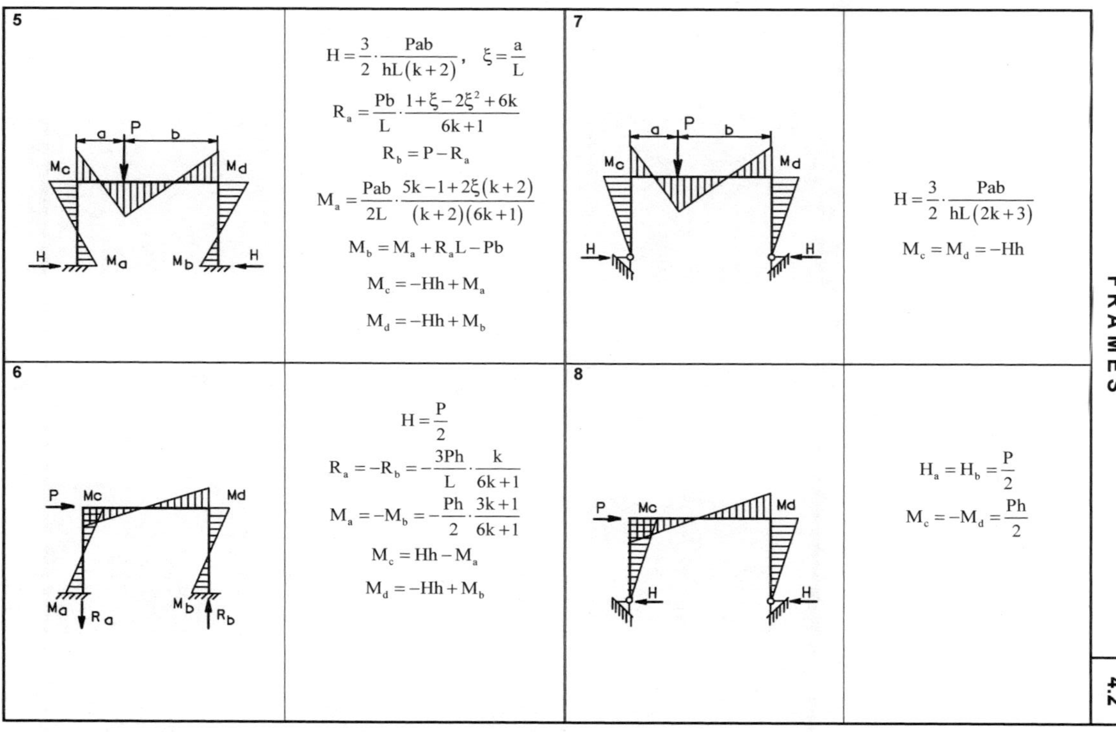

5

$$H=\frac{3}{2}\cdot\frac{Pab}{hL(k+2)},\quad \xi=\frac{a}{L}$$

$$R_a=\frac{Pb}{L}\cdot\frac{1+\xi-2\xi^2+6k}{6k+1}$$

$$R_b=P-R_a$$

$$M_a=\frac{Pab}{2L}\cdot\frac{5k-1+2\xi(k+2)}{(k+2)(6k+1)}$$

$$M_b=M_a+R_aL-Pb$$

$$M_c=-Hh+M_a$$

$$M_d=-Hh+M_b$$

6

$$H=\frac{P}{2}$$

$$R_a=-R_b=-\frac{3Ph}{L}\cdot\frac{k}{6k+1}$$

$$M_a=-M_b=-\frac{Ph}{2}\cdot\frac{3k+1}{6k+1}$$

$$M_c=Hh-M_a$$

$$M_d=-Hh+M_b$$

7

$$H=\frac{3}{2}\cdot\frac{Pab}{hL(2k+3)}$$

$$M_c=M_d=-Hh$$

8

$$H_a=H_b=\frac{P}{2}$$

$$M_c=-M_d=\frac{Ph}{2}$$

NOTES

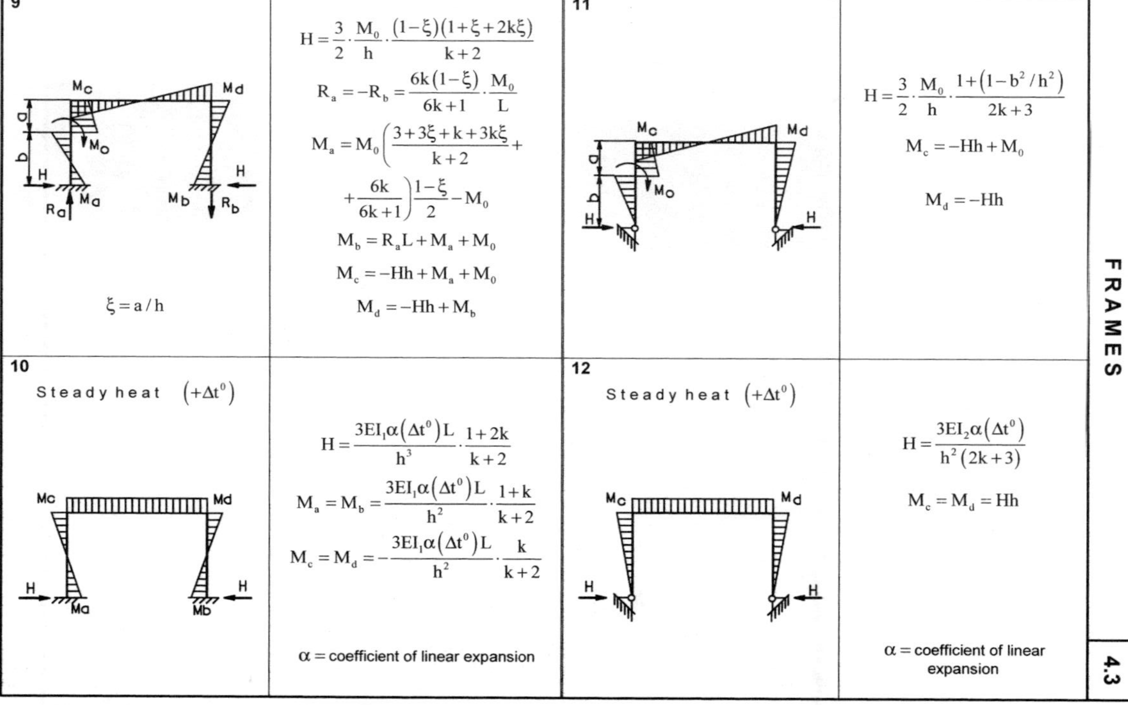

9

$$H = \frac{3}{2}\cdot\frac{M_0}{h}\cdot\frac{(1-\xi)(1+\xi+2k\xi)}{k+2}$$

$$R_a = -R_b = \frac{6k(1-\xi)}{6k+1}\cdot\frac{M_0}{L}$$

$$M_a = M_0\left(\frac{3+3\xi+k+3k\xi}{k+2} + \frac{6k}{6k+1}\right)\frac{1-\xi}{2} - M_0$$

$$M_b = R_a L + M_a + M_0$$

$$M_c = -Hh + M_a + M_0$$

$$M_d = -Hh + M_b$$

$$\xi = a/h$$

10 Steady heat $(+\Delta t^0)$

$$H = \frac{3EI_1\alpha(\Delta t^0)L}{h^3}\cdot\frac{1+2k}{k+2}$$

$$M_a = M_b = \frac{3EI_1\alpha(\Delta t^0)L}{h^2}\cdot\frac{1+k}{k+2}$$

$$M_c = M_d = -\frac{3EI_1\alpha(\Delta t^0)L}{h^2}\cdot\frac{k}{k+2}$$

α = coefficient of linear expansion

11

$$H = \frac{3}{2}\cdot\frac{M_0}{h}\cdot\frac{1+(1-b^2/h^2)}{2k+3}$$

$$M_c = -Hh + M_0$$

$$M_d = -Hh$$

12 Steady heat $(+\Delta t^0)$

$$H = \frac{3EI_2\alpha(\Delta t^0)}{h^2(2k+3)}$$

$$M_c = M_d = Hh$$

α = coefficient of linear expansion

NOTES

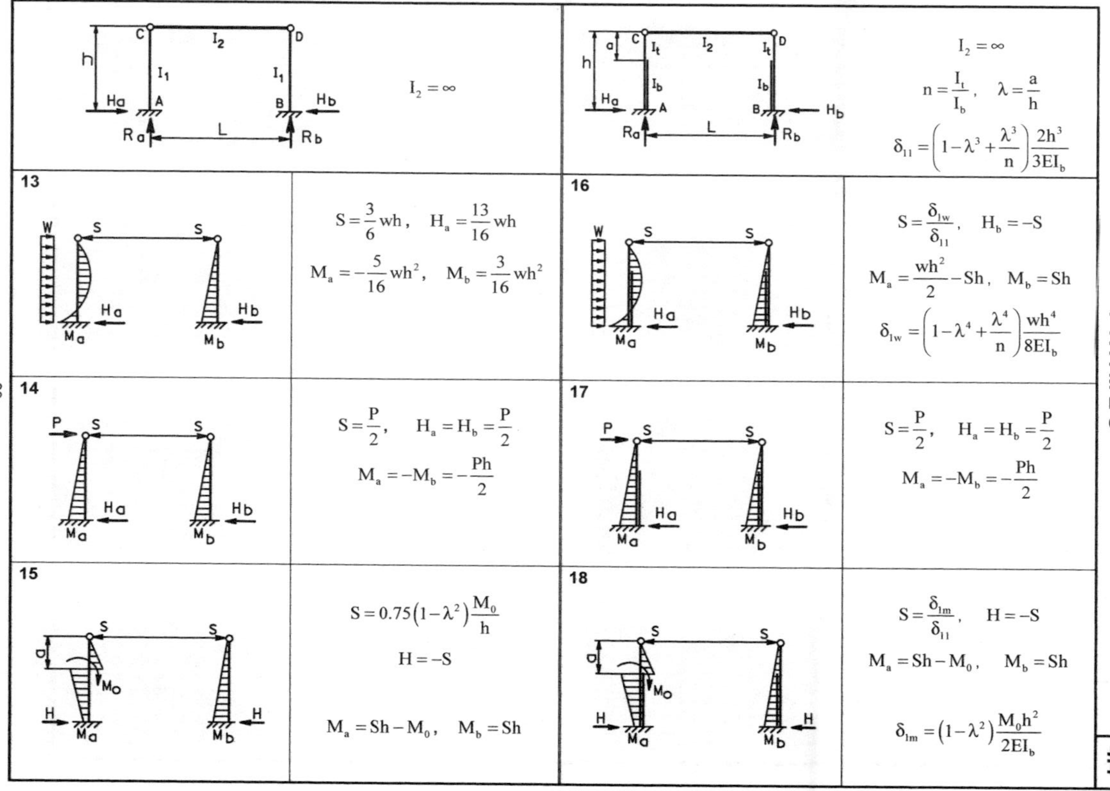

$I_2 = \infty$

$I_2 = \infty$

$n = \frac{I_t}{I_b}, \quad \lambda = \frac{a}{h}$

$\delta_{11} = \left(1 - \lambda^3 + \frac{\lambda^3}{n}\right)\frac{2h^3}{3EI_b}$

13

$S = \frac{3}{6}wh, \quad H_a = \frac{13}{16}wh$

$M_a = -\frac{5}{16}wh^2, \quad M_b = \frac{3}{16}wh^2$

14

$S = \frac{P}{2}, \quad H_a = H_b = \frac{P}{2}$

$M_a = -M_b = -\frac{Ph}{2}$

15

$S = 0.75\left(1 - \lambda^2\right)\frac{M_0}{h}$

$H = -S$

$M_a = Sh - M_0, \quad M_b = Sh$

16

$S = \frac{\delta_{1w}}{\delta_{11}}, \quad H_b = -S$

$M_a = \frac{wh^2}{2} - Sh, \quad M_b = Sh$

$\delta_{1w} = \left(1 - \lambda^4 + \frac{\lambda^4}{n}\right)\frac{wh^4}{8EI_b}$

17

$S = \frac{P}{2}, \quad H_a = H_b = \frac{P}{2}$

$M_a = -M_b = -\frac{Ph}{2}$

18

$S = \frac{\delta_{1m}}{\delta_{11}}, \quad H = -S$

$M_a = Sh - M_0, \quad M_b = Sh$

$\delta_{1m} = \left(1 - \lambda^2\right)\frac{M_0 h^2}{2EI_b}$

NOTES

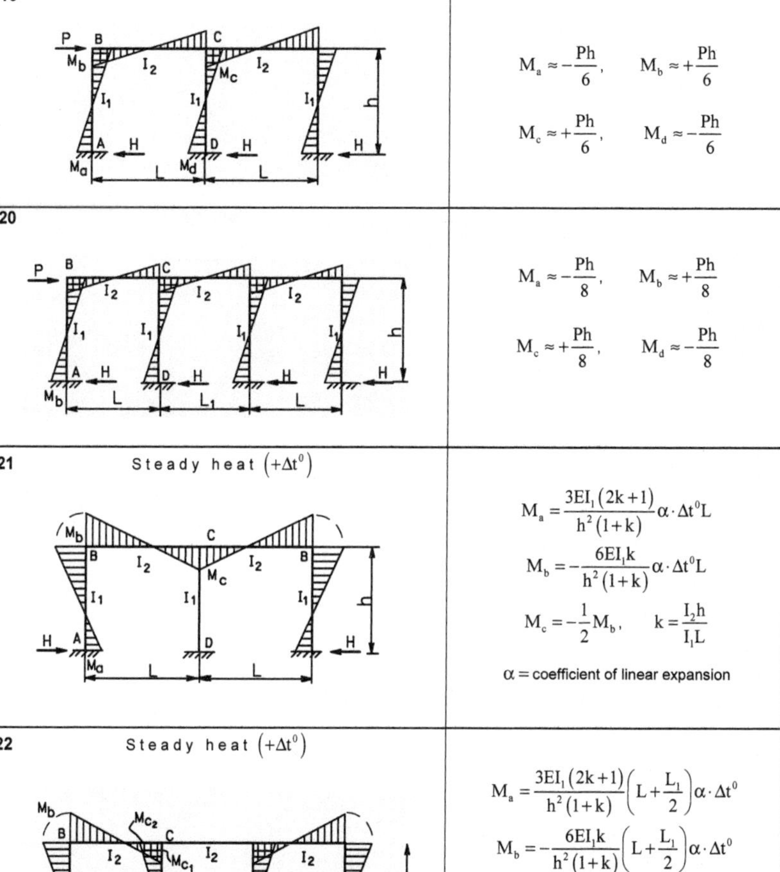

19

$$M_a \approx -\frac{Ph}{6}, \qquad M_b \approx +\frac{Ph}{6}$$

$$M_c \approx +\frac{Ph}{6}, \qquad M_d \approx -\frac{Ph}{6}$$

20

$$M_a \approx -\frac{Ph}{8}, \qquad M_b \approx +\frac{Ph}{8}$$

$$M_c \approx +\frac{Ph}{8}, \qquad M_d \approx -\frac{Ph}{8}$$

21 Steady heat $(+\Delta t^0)$

$$M_a = \frac{3EI_1(2k+1)}{h^2(1+k)}\alpha \cdot \Delta t^0 L$$

$$M_b = -\frac{6EI_1 k}{h^2(1+k)}\alpha \cdot \Delta t^0 L$$

$$M_c = -\frac{1}{2}M_b, \qquad k = \frac{I_2 h}{I_1 L}$$

α = coefficient of linear expansion

22 Steady heat $(+\Delta t^0)$

Mb, B, Mc2, C, Mc1, I2, I2, I2, I1, I1, I1, I1, h, H, A, H, D, H, H, Mb, L, L1, L, a

$$M_a = \frac{3EI_1(2k+1)}{h^2(1+k)}\left(L+\frac{L_1}{2}\right)\alpha \cdot \Delta t^0$$

$$M_b = -\frac{6EI_1 k}{h^2(1+k)}\left(L+\frac{L_1}{2}\right)\alpha \cdot \Delta t^0$$

$$M_{c1} = -\frac{1}{2}M_b, \quad M_{c2} = -\frac{6EI_1}{h^2}\left(\frac{L_1}{2}\right)\alpha\Delta t^0$$

$$M_d = -M_{c2}, \quad k = \frac{I_2 h}{I_1 L}$$

α = coefficient of linear expansion

NOTES

5. ARCHES

Diagrams and Formulas for Various Loading Conditions

NOTES

Tables 5.1–5.9 are provided for determining support reactions and bending moments in elastic arches with constant or variable cross-sections.

Table 5.1 includes formulas for computing in any cross-section k the axis force N_k and the shear V_k. These formulas can also be applied in analysis of arches shown in Tables 5.2–5.9.

Bending moment $$M_k = R_A \cdot x_k - H_A \cdot y_k \pm M_A - \sum_{\text{Left}} P_i \cdot a_i$$

Axial force $$N_k = R_A \sin\phi + H_A \cos\phi - \sum_{\text{Left}} P_i \sin\phi$$

Shear $$V_k = R_A \cos\phi - H_A \sin\phi - \sum_{\text{Left}} P_i \cos\phi$$

Where a_i = distance from load P to point k.

THREE-HINGED ARCHES

SUPPORT REACTIONS, BENDING MOMENT and AXIAL FORCE

5.1

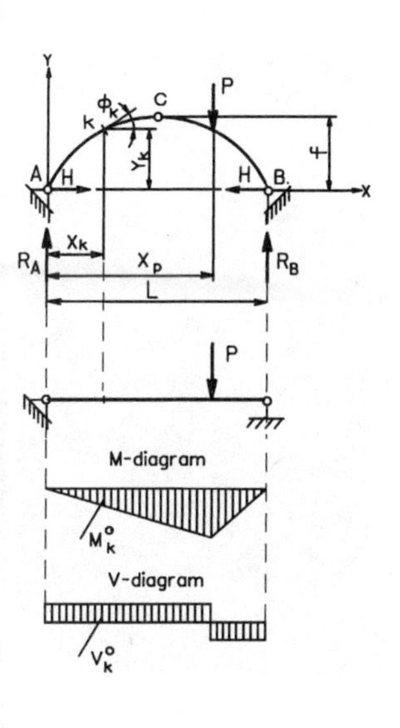

Vertical reactions:

$$\sum M_B = R_A L - P(L - x_P) = 0, \quad R_A = P\frac{L - x_P}{L};$$

$$\sum M_A = -R_B L + P x_P = 0, \quad R_B = P\frac{x_P}{L}.$$

Horizontal reactions:

$$\sum_{Left} M_C = R_A\frac{L}{2} - H_A f = 0, \quad H_A = R_A\frac{L}{2f};$$

$$\sum X = H_A - H_B = 0, \quad H_B = H_A = H.$$

Section k (x_k, y_k)

Bending moment: $M_k = \sum_{Left} M = R_A x_k - H y_k,$

or $M_k = M_k^0 - H y_k$.

Shear: $V_k = \left(R_A - \sum_{Left} P\right)\cos\phi_k - H\sin\phi_k$

or $V_k = V_k^0 \cos\phi_k - H\sin\phi_k$.

Axial force: $N_k = \left(R_A - \sum_{Left} P\right)\sin\phi_k + H\cos\phi_k$

or $N_k = V_k^0 \sin\phi_k + H\cos\phi_k$.

M_k^0 and V_k^0 = bending moment and shear in simple beam for section x_k

Tied arch

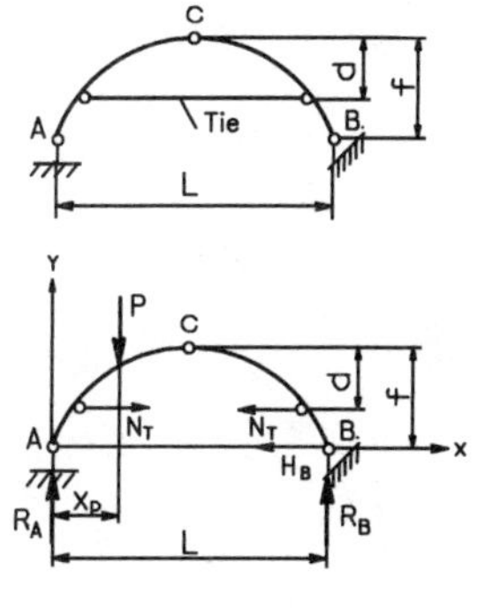

Vertical reactions:

$$\sum M_B = R_A L - P(L - x_P) = 0, \quad R_A = P\frac{L - x_P}{L};$$

$$\sum M_A = -R_B L + P x_P = 0, \quad R_B = P\frac{x_P}{L}.$$

Horizontal reaction:

$$\sum X = -H_B = 0.$$

Force N_T:

$$\sum_{Left} M_C = R_A\frac{L}{2} - N_T d - P\left(\frac{L}{2} - x_P\right) = 0,$$

$$N_T = \frac{1}{d}\left[P\left(\frac{L}{2} - x_P\right) - R_A\frac{L}{2}\right]$$

or $\sum_{Right} M_C = N_T d - R_B\frac{L}{2} = 0, \quad N_T = R_b\frac{L}{2d}.$

NOTES

Table 5.2

Example. Symmetrical three-hinged arch

Given. Circular arch 2 in Table 5.2, $L = 20$ m, $f = 4$ m,

radius $R = \dfrac{4f^2 + L^2}{8f} = \dfrac{4\times 4^2 + 20^2}{8\times 4} = 14.5$ m, $x_m = 5$ m,

$$y_m = \sqrt{R^2 - \left(\frac{L}{2} - x_m\right)^2} - (R - f) = \sqrt{14.5^2 - (10-5)^2} - (14.5 - 4) = 3.11 \text{ m}$$

$$\tan\phi_m = \left(\frac{L}{2} - x_m\right) / (R - f + y_m) = (10-5)/(14.5 - 4 + 3.11) = 0.367$$

$\phi_m = 20.17^0$, $\sin\phi_m = 0.345$, $\cos\phi_m = 0.939$

Distribution load $w = 2$ kN/m

Required. Compute support reactions R_A and H_A, support bending moment M_A, bending moment M_m, axial force N_m and shear V_m

Solution. $R_A = \dfrac{3}{8} wL = \dfrac{3}{8}\times 2\times 20 = 15$ kN, $H_A = \dfrac{wL^2}{16f} = \dfrac{2\times 20^2}{16\times 4} = 12.5$ kN

$$\xi_m = \frac{x_m}{L} = \frac{5}{20} = 0.25, \quad \eta_m = \frac{y_m}{f} = \frac{3.11}{4} = 0.778$$

$$M_m = \frac{wL^2}{16}\left[8(\xi_m - \xi_m^2) - 2\xi_m - \eta_m\right] = \frac{2\times 20^2}{16}\left[8(0.25 - 0.25^2) - 2\times 0.25 - 0.778\right] = 11.1 \text{ kN}\cdot\text{m}$$

$$N_m = R_A \sin\phi_m + H_A \cos\phi_m - w\cdot x_m \sin\phi_m = 15\times 0.345 + 12.5\times 0.939 - 2\times 5\times 0.345 = 13.46 \text{ kN}$$

$$V_m = R_A \cos\phi_m - H_A \sin\phi_m - w\cdot x_m \cos\phi_m = 15\times 0.939 - 12.5\times 0.345 - 2\times 5\times 0.939 = 0.38 \text{ kN}$$

SYMMETRICAL THREE–HINGED ARCHES OF ANY SHAPE

FORMULAS for VARIOUS STATIC LOADING CONDITIONS

5.2

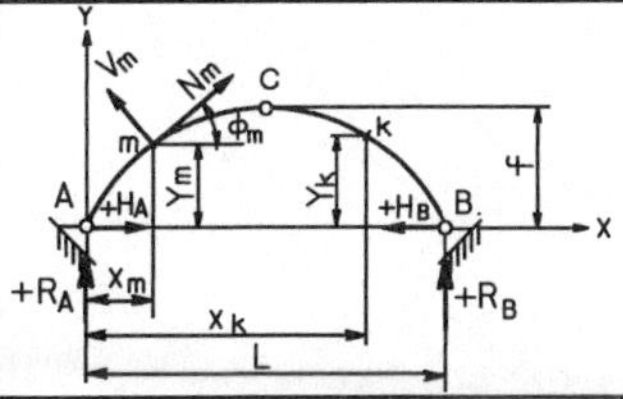

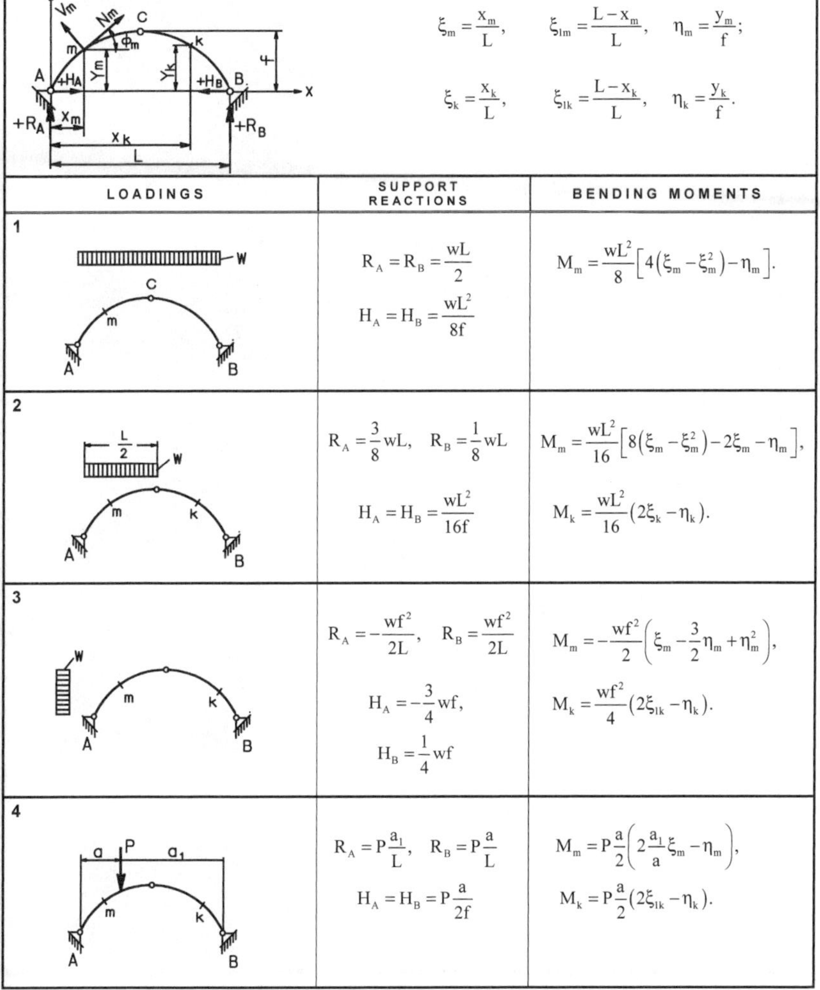

$$\xi_m = \frac{x_m}{L}, \qquad \xi_{1m} = \frac{L - x_m}{L}, \qquad \eta_m = \frac{y_m}{f};$$

$$\xi_k = \frac{x_k}{L}, \qquad \xi_{1k} = \frac{L - x_k}{L}, \qquad \eta_k = \frac{y_k}{f}.$$

LOADINGS	SUPPORT REACTIONS	BENDING MOMENTS
1 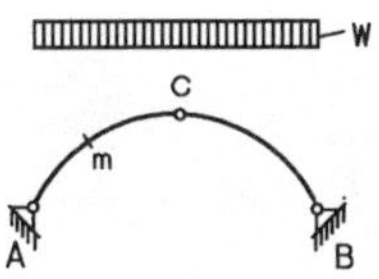	$R_A = R_B = \frac{wL}{2}$ $H_A = H_B = \frac{wL^2}{8f}$	$M_m = \frac{wL^2}{8}\left[4\left(\xi_m - \xi_m^2\right) - \eta_m\right].$
2 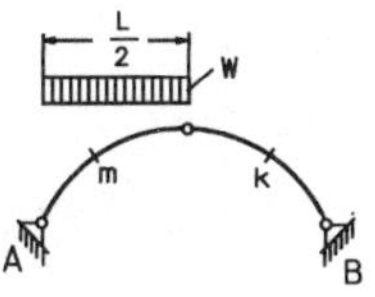	$R_A = \frac{3}{8}wL, \quad R_B = \frac{1}{8}wL$ $H_A = H_B = \frac{wL^2}{16f}$	$M_m = \frac{wL^2}{16}\left[8\left(\xi_m - \xi_m^2\right) - 2\xi_m - \eta_m\right],$ $M_k = \frac{wL^2}{16}\left(2\xi_k - \eta_k\right).$
3 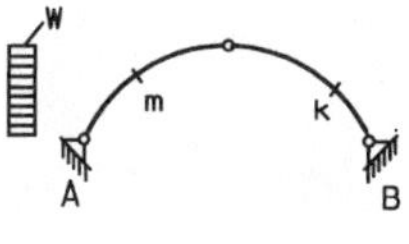	$R_A = -\frac{wf^2}{2L}, \quad R_B = \frac{wf^2}{2L}$ $H_A = -\frac{3}{4}wf,$ $H_B = \frac{1}{4}wf$	$M_m = -\frac{wf^2}{2}\left(\xi_m - \frac{3}{2}\eta_m + \eta_m^2\right),$ $M_k = \frac{wf^2}{4}\left(2\xi_{1k} - \eta_k\right).$
4 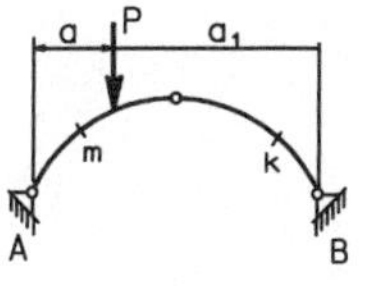	$R_A = P\frac{a_1}{L}, \quad R_B = P\frac{a}{L}$ $H_A = H_B = P\frac{a}{2f}$	$M_m = P\frac{a}{2}\left(2\frac{a_1}{a}\xi_m - \eta_m\right),$ $M_k = P\frac{a}{2}\left(2\xi_{1k} - \eta_k\right).$

NOTES

SYMMETRICAL THREE-HINGED ARCHES OF ANY SHAPE

FORMULAS for VARIOUS STATIC LOADING CONDITIONS

5.3

LOADINGS	SUPPORT REACTIONS	BENDING MOMENTS
5 w, m, k, A, B	$R_A = \frac{5}{24} wL$ $R_B = \frac{1}{24} wL$ $H_A = H_B = \frac{wL^2}{48f}$	$M_m = \frac{wL^2}{48}\left[2\xi_m + 8\left(\xi_{1m} - \xi_{1m}^3 - \xi_m + \xi_m^3\right) - \eta_m\right],$ $M_k = \frac{wL^2}{48}\left(2\xi_{1k} - \eta_k\right).$
6 w, w, m, A	$R_A = R_B = \frac{wL}{4}$ $H_A = H_B = \frac{wL^2}{24f}$	$M_m = \frac{wL^2}{24}\left[2\xi_m + 4\left(\xi_{1m} - \xi_{1m}^3 - \xi_m + \xi_m^3\right) - \eta_m\right],$
7 w, m, k, A, B	$R_A = -\frac{wf^2}{6L},\ R_B = \frac{wf^2}{6L}$ $H_A = -\frac{5}{12} wf,$ $H_B = \frac{1}{12} wf$	$M_m = \frac{wL^2}{12}\left[2\left(\xi_{1m} - \xi_{1m}^3\right) + \eta_m - 2\xi_m\right],$ $M_k = \frac{wL^2}{12}\left(2\xi_{1k} - \eta_k\right).$
8 M_0, M_0, m, A, B	$R_A = R_B = 0$ $H_A = H_B = -\frac{M_0}{f}$	$M_m = M_0 \eta_m$

N O T E S

Table 5.4

Example. Two-hinged parabolic arch

Given. Parabolic arch 3 in Table 5.4

$$L = 20 \text{ m},\ f = 3 \text{ m},\ x = a = 5 \text{ m},\ \xi = \frac{a}{L} = \frac{5}{20} = 0.25$$

$$\tan\phi_x = \frac{4f(L-2x)}{L^2} = \frac{4\times3(20-2\times5)}{20^2} = 0.3,$$

$$\phi_x = 16.7^0,\ \sin\phi_x = 0.287,\ \cos\phi_x = 0.958$$

Concentrated load $P = 20 \text{ kN}$

Required. Compute support reactions R_A and H_A, bending moments M_c and M_x, axial force N_x and shear V_x (at point of load)

Solution. $R_A = P\frac{L-a}{L} = 20\frac{20-5}{20} = 15 \text{ kN}$

$$H_A = \frac{5PL}{8f}k\left[\xi - 2\xi^2 + \xi^4\right] = \frac{5\times20\times20}{8\times3}\times1\times\left[0.25 - 2\times0.25^2 + 0.25^4\right] = 10.75 \text{ kN}$$

$$M_c = \frac{PL}{8}\left[4\xi - 5k\left(\xi - 2\xi^3 + \xi^4\right)\right] = \frac{20\times20}{8}\left[4\times0.25 - 5\left(0.25 - 2\times0.25^3 + 0.25^4\right) = -9.5 \text{ kN}\cdot\text{m}\right]$$

$$y_x = \frac{4f(L-x)x}{L^2} = \frac{4\times3(20-5)\times5}{20^2} = 2.25$$

$$M_x = R_A a - H_A y_x = 15\times5 - 10.75\times2.25 = 50.81 \text{ kN}\cdot\text{m}$$

$$N_x = R_A\sin\phi_x + H_A\cos\phi_x = 15\times0.287 + 10.75\times0.958 = 14.6 \text{ kN}$$

$$V_x = R_A\cos\phi_x - H_A\sin\phi_x = 15\times0.958 - 10.75\times0.287 = 11.3 \text{ kN}$$

TWO–HINGED PARABOLIC ARCHES

FORMULAS for VARIOUS STATIC LOADING CONDITIONS

5.4

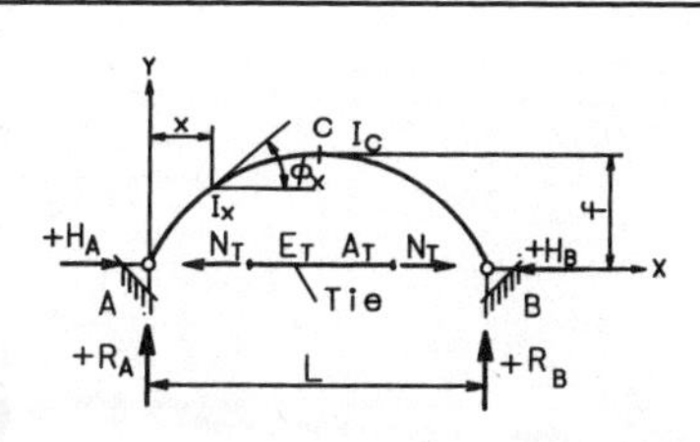

Equation of parabola:

$$y=\frac{4f(L-x)x}{L^2},\quad I_x=I_c/\cos\phi_x$$

$$\tan\phi=\frac{dy}{dx}=\frac{4f(L-2x)}{L^2}$$

Coefficients: For regular arch: $\upsilon=0,\quad k=1$

For tied arch: $\upsilon=\frac{15}{8}\cdot\frac{\beta}{f^2},\quad k=\frac{1}{1+\upsilon},\quad \beta=\frac{EI_c}{E_T A_T}$

LOADINGS	SUPPORT REACTIONS	BENDING MOMENTS
1 (w; C; A; B)	$R_A=R_B=\frac{wL}{2}$ $H_A=H_B=\frac{wL^2}{8f}k$	$M_C=\frac{wL^2}{8}(1-k)$ $\upsilon=\frac{15}{8}\cdot\frac{\beta}{f^2},\quad k=\frac{1}{1+\upsilon}.$
2 ($\frac{L}{2}$; w; m; C; A; B; $\frac{L}{4}$)	$R_A=\frac{3}{8}wL,\quad R_B=\frac{1}{8}wL$ $H_A=H_B=\frac{wL^2}{16f}k$	$M_C=\frac{wL^2}{16}(1-k),$ $M_m=\left(\frac{1}{16}-\frac{3}{64}k\right)wL^2$
3 ($a=\xi L$; P; C; A; B)	$R_A=P\frac{L-a}{L},\quad R_B=P\frac{a}{L}$ $H_A=H_B$ $=\frac{5PL}{8f}k\left[\xi-2\xi^3+\xi^4\right]$	$M_C=\frac{PL}{8}\left[4\xi-5k\left(\xi-2\xi^3+\xi^4\right)\right],$ $\xi=\frac{a}{L}.$
4 (w; C; A; B)	$R_A=\frac{5wL}{24},\quad R_B=\frac{wL}{24}$ $H_A=H_B=0.0228\frac{wL^2}{f}k$	$M_C=R_B\frac{L}{2}-H_B f$

NOTES

TWO-HINGED PARABOLIC ARCHES

FORMULAS for VARIOUS STATIC LOADING CONDITIONS

5.5

LOADINGS	SUPPORT REACTIONS	BENDING MOMENTS
5 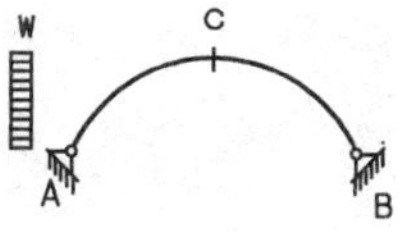	$R_A = -\frac{wf^2}{2L}, \quad R_B = -R_A$ $H_A = -0.714wf$ $H_B = 0.286wf$	$M_C = -0.0357wf^2$
6 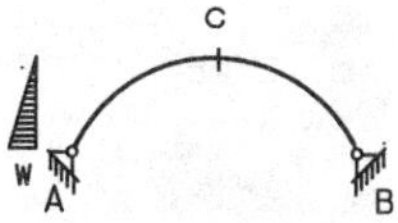	$R_A = -\frac{wf^2}{6L}, \quad R_B = -R_A$ $H_A = -0.401wf$ $H_B = 0.099wf$	$M_C = -0.0159wf^2$
7 Tied arch 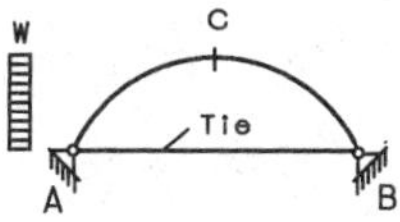	$R_A = -\frac{wf^2}{2L}, \quad R_B = -R_A$ $H = wf$ $N_T = \frac{2.286wf^3}{8f^2 + 15\beta}$	$M_C = \frac{wf^2}{4} - N_T f$
8 Tied arch 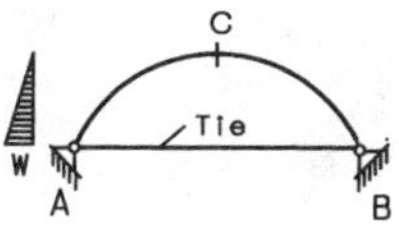	$R_A = -\frac{wf^2}{6L}, \quad R_B = -R_A$ $H = \frac{wf}{2}$ $N_T = \frac{0.792wf^3}{8f^2 + 15\beta}$	$M_C = \frac{wf^2}{12} - N_T f$
9 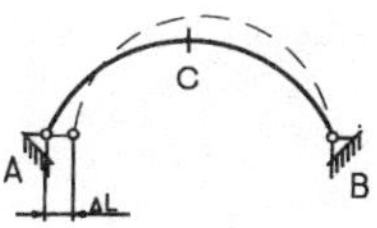	$R_A = R_B = 0$ $H = \frac{15}{8} \cdot \frac{EI_C \Delta_L}{f^2 L} k$	$M_C = -Hf$

NOTES

Table 5.6

Example. Fixed parabolic arch

Given. Fixed parabolic arch 2 in Table 5.6

$$L = 20 \text{ m}, \ f = 3 \text{ m}, \ x = \xi L = 8 \text{ m}, \ \xi = \frac{8}{20} = 0.4, \ \xi_1 = \frac{L-x}{L} = \frac{20-8}{20} = 0.6$$

Distribution load $w = 2$ kN/m

Required. Compute support reactions R_A and H_A, bending moments M_A and M_C

Solution.

$$R_A = \frac{wL}{2}\xi\left[1+\xi_1(1+\xi\xi_1)\right] = \frac{2\times 20}{2}0.4\left[1+0.6(1+0.4\times 0.6)\right] = 13.95 \text{ kN}$$

$$H_A = \frac{wL^2}{8f}\xi^3\left[1+3\xi_1(1+2\xi_1)\right] = \frac{2\times 20^2}{8\times 3}\times 0.4^3\times\left[1+3\times 0.6(1+2\times 0.6)\right] = 10.58 \text{ kN}$$

$$M_A = -\frac{wL^2}{2}\xi^2\xi_1^3 = -\frac{2\times 20^2}{2}\times 0.4^2\times 0.6^3 = -13.82 \text{ kN}\cdot\text{m}$$

$$M_C = R_A\frac{L}{2} - w\times 8\times 6 - H_A f - M_A$$

$$= 13.95\times 10 - 2\times 8\times 6 - 10.58\times 3 - 13.82 = -2.06 \text{ kN}\cdot\text{m}$$

FIXED PARABOLIC ARCHES

FORMULAS for VARIOUS STATIC LOADING CONDITIONS

5.6

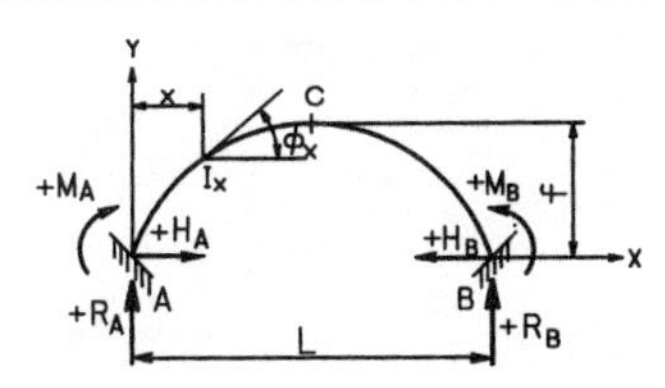

Equation of parabola:

$$y = \frac{4f(L-x)x}{L^2}, \quad I_x = I_C / \cos\phi_x$$

$$\tan\phi = \frac{dy}{dx} = \frac{4f(L-2x)}{L^2}$$

$$\xi = \frac{x}{L}, \quad \xi_1 = \frac{L-x}{L}$$

LOADINGS	SUPPORT REACTIONS	BENDING MOMENTS
1	$R_A = R_B = \frac{wL}{2}$ $H_A = H_B = \frac{wL^2}{8f}k$	$M_A = M_B = M_C = 0$
2 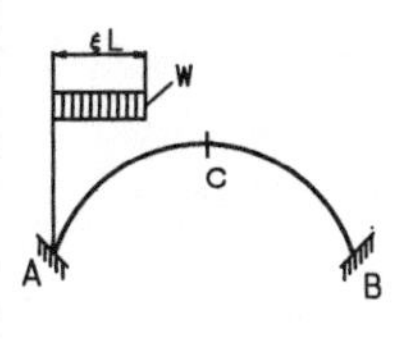	$R_A = \frac{wL}{2}\xi\left[1+\xi_1(1+\xi\xi_1)\right]$ $R_B = \frac{wL}{2}\xi^2(1-\xi_1^2)$ $H = \frac{wL^2}{8f}\xi^3\left[1+3\xi_1(1+2\xi_1)\right]$	$M_A = -\frac{wL^2}{2}\xi^2\xi_1^3$ $M_B = \frac{wL^2}{2}\xi^3\xi_1^2$
3	$R_A = -\frac{wf^2}{4L}, \quad R_B = \frac{wf^2}{4L}$ $H_A = -\frac{11}{14}wf$ $H_B = \frac{3}{14}wf$	$M_A = -\frac{51}{280}wf^2$ $M_B = \frac{19}{280}wf^2$ $M_C = -\frac{3}{140}wf^2$
4 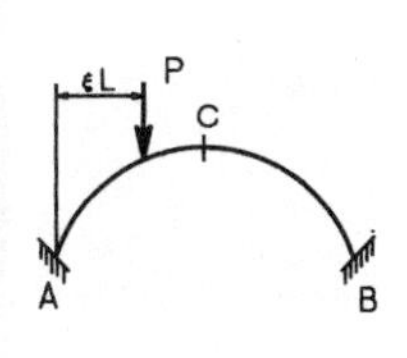	$R_A = \xi_1^2(1+2\xi)P$ $R_B = \xi^2(1+2\xi_1)P$ $H = P\frac{15L}{4f}\xi^2\xi_1^2$	$M_A = PL\xi\xi_1^2\left(\frac{5}{2}\xi-1\right)$ $M_B = PL\xi^2\xi_1\left(\frac{5}{2}\xi_1-1\right)$ For $0 \le \xi \le 0.5$: $M_C = \frac{PL}{2}\xi^2\left(1-\frac{5}{2}\xi_1^2\right)$

NOTES

FIXED PARABOLIC ARCHES

FORMULAS for VARIOUS STATIC LOADING CONDITIONS

5.7

LOADINGS	SUPPORT REACTIONS	BENDING MOMENTS
5	$R_A = R_B = \frac{P}{2}$ $H = \frac{15PL}{64f}$	$M_A = M_B = \frac{PL}{32}$ $M_C = \frac{3PL}{64}$
6 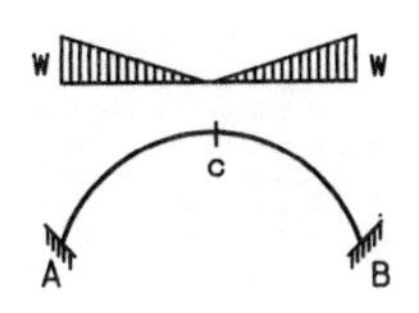	$R_A = R_B = \frac{wL}{4}$ $H = \frac{5wL^2}{128f}$	$M_A = M_B = -\frac{wL^2}{192}$ $M_C = -\frac{wL^2}{384}$
7 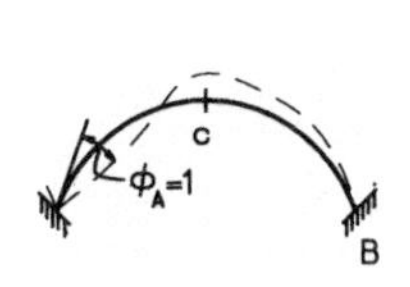	$R_A = -\frac{6EI_C}{L^2}$ $R_B = +\frac{6EI_C}{L^2}$ $H = \frac{15}{2f} \cdot \frac{EI_C}{L}$	$M_A = \frac{9EI_C}{L}$ $M_B = \frac{3EI_C}{L}$ $M_C = -\frac{3}{2} \cdot \frac{EI_C}{L}$
8 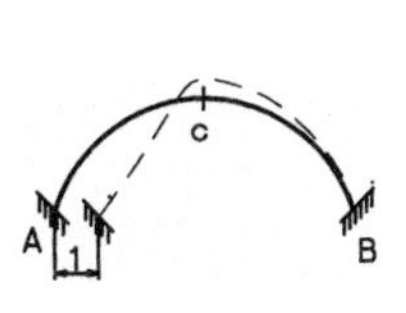	$R_A = R_B = 0$ $H = \frac{45}{4} \cdot \frac{EI_C}{f^2 L}$	$M_A = M_B = \frac{15}{2f} \cdot \frac{EI_C}{L}$ $M_C = -\frac{15}{4f} \cdot \frac{EI_C}{L}$

NOTES

THREE-HINGED ARCHES

INFLUENCE LINES

5.8

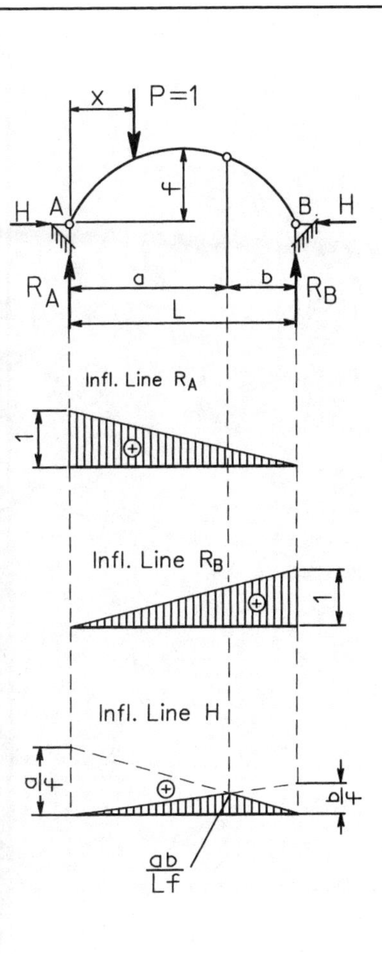

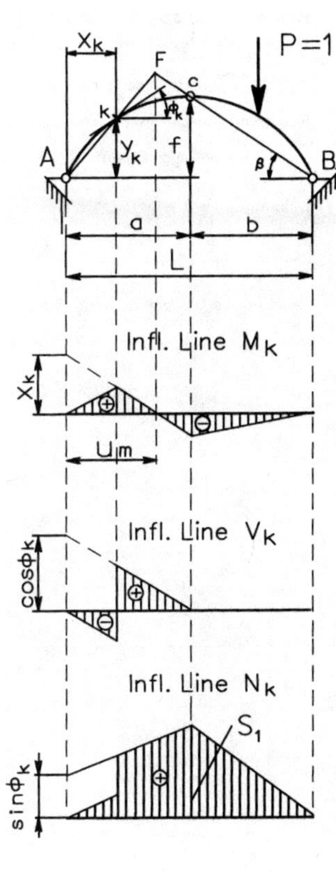

$$u_m = \frac{L \cdot f \cdot x_k}{y_k \cdot b + x_k \cdot f},$$

$$S_1 = \frac{a - u_n}{-u_n} \sin \phi_k, \quad u_n = \frac{L \cdot \tan \beta}{\tan \beta - \cot \phi_k}.$$

N O T E S

Table 5.9

Example. Fixed parabolic arch

Given. $L = 40$ m, $f = 10$ m, $x_k = 8$ m

Concentrated load in point k $P_k = 12$ kN

Required. Using influence lines, compute support reactions R_A and H_A, support bending moment M_A, bending moments M_c and M_k, axial force N_k, and shear V_k

Solution. $\frac{x_k}{L} = \frac{8}{40} = 0.2, \quad y_k = \frac{4 \times 10(40-8)8}{40^2} = 6.4 \text{ m},$

$\tan\phi = \frac{4f(L-2x)}{L^2} = \frac{4 \times 10(40 - 2 \times 8)}{40^2} = 0.6,$

$\phi_k = 30.96^0, \quad \sin\phi_k = 0.514, \quad \cos\phi_k = 0.857$

$$R_A = S_i \times P_k = 0.896 \times 12 = 10.752 \text{ kN}$$

$$H_A = S_i \times \frac{L}{f} \times P_k = 0.0960 \times \frac{40}{10} \times 12 = 4.608 \text{ kN}$$

$$M_A = S_i \times L \times P_k = -0.0640 \times 40 \times 12 = 30.72 \text{ kN} \cdot \text{m}$$

$$M_c = S_i \times L \times P_k = -0.0120 \times 40 \times 12 = -5.76 \text{ kN} \cdot \text{m}$$

$$M_k = R_A \cdot x_k - H_A \cdot y_k - M_A = 10.752 \times 8 - 4.608 \times 6.4 - 30.72 = 25.805 \text{ kN} \cdot \text{m}$$

$$N_k = R_A \sin\phi_k + H_A \cos\phi_k = 10.752 \times 0.514 + 4.608 \times 0.857 = 9.475 \text{ kN}$$

$$V_k = R_A \cos\phi_k - H_A \sin\phi_k = 10.752 \times 0.857 - 4.608 \times 0.514 = 6.745 \text{ kN}$$

FIXED PARABOLIC ARCHES

INFLUENCE LINES

5.9

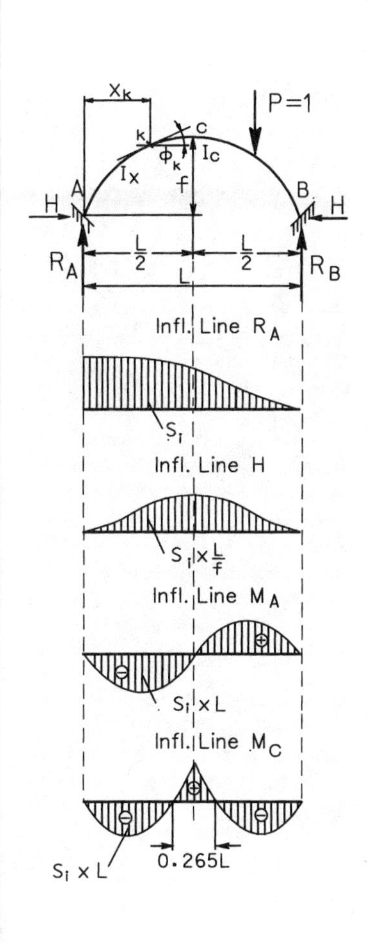

Equation of parabola: $y = \dfrac{4f(L-x)x}{L^2}$

$I_x = I_C / \cos\phi_x$, $\tan\phi = \dfrac{dx}{dy} = \dfrac{4f(L-2x)}{L^2}$

$R_A = S_i \times P$, $H = S_i \times \dfrac{L}{f} \times P$, $M = S_i \times L \times P$

ORDINATES OF INFLUENCE LINES (S_i)

$\frac{x}{L}$	R_A	H	M_A	M_C
0.0	1.000	0.0	0.0	0.0
0.05	0.993	0.0085	–0.0395	–0.0016
0.10	0.972	0.0305	–0.0625	–0.0052
0.15	0.939	0.0610	–0.0678	–0.0090
0.20	0.896	0.0960	–0.0640	–0.0120
0.25	0.844	0.1320	–0.0528	–0.0127
0.30	0.784	0.1655	–0.0368	–0.0102
0.35	0.718	0.1940	–0.0184	–0.0034
0.40	0.648	0.2160	0.0	0.0080
0.45	0.575	0.2295	0.0174	0.0246
0.50	0.500	0.2344	0.0312	0.0468
0.55	0.425	0.2295	0.0418	0.0246
0.60	0.352	0.2160	0.0480	0.0080
0.65	0.282	0.1940	0.0498	–0.0034
0.70	0.216	0.1655	0.0473	–0,0102
0.75	0.156	0.1320	0.0410	–0,0127
0.80	0.104	0.0960	0.0320	–0.0120
0.85	0.061	0.0610	0.0215	–0.0090
0.90	0.028	0.0305	0.0118	–0,0052
0.95	0.007	0.0085	0.0032	–0.0016
1.00	0.0	0.0	0.0	0,0

NOTES

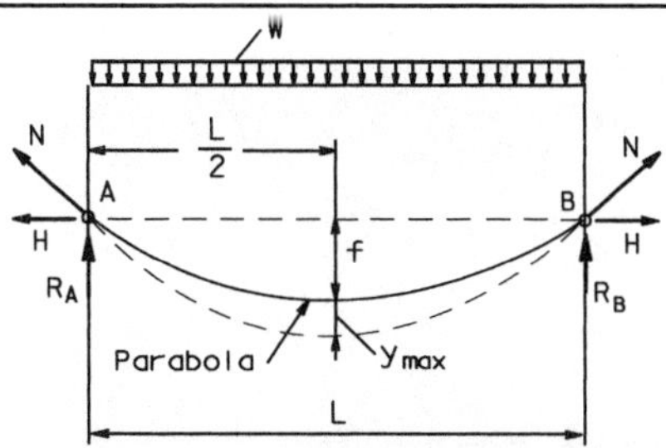

Rope deflection

w = uniformly distributed load, f = rope sag due to natural weight, $\left(f \approx 1/20 \cdot L\right)$

s = length of rope, $s = \sqrt{L^2 + \frac{16}{3} f^2}$

Forces and deflection:

$H = \frac{\sqrt{0.25 w L^4}}{4f}$ (elastic deformations are not included)

$H = \sqrt[3]{\frac{w^2 L^2 EA}{24}}$ (elastic deformations are included)

E = modulus of elasticity, A = area of rope cross-section

$N_{max} = \sqrt{H^2 + R^2}$, R = reaction, $R = wL/2$

Bending moment $M_{max} = wL^2/8$, Deflection $y_{max} = \frac{M_{max}}{H}$

Temperature:

$N_t = \alpha \cdot \Delta t^0 \cdot EA$, $\Delta t^0 = T_1^0 - T_2^0$, if: $\Delta t^0 > 0$ (tension), $\Delta t^0 < 0$ (compression)

α = linear coefficient of expansion

$H_t^3 - N_t \cdot H_t^2 = \frac{wL^2 EA}{24}$ $H_t^3 - N_t H_t^2 = \frac{wL^2 EA}{24}$, $N_{max} = \sqrt{H_t^2 + R^2}$, $y_{max} = \frac{M_{max}}{H_t}$

NOTES

6. TRUSSES

Method of Joints
and
Method of Section Analysis

N O T E S

Tables 6.1–6.4 provide examples of analysis of flat trusses.

Legend

- Upper chord: U
- Lower chord: L
- Vertical posts: $U_i - L_i$
- Diagonals: $U_i - L_{i\pm1}$
- End posts: $L_0 - U_1$
- Load on upper chord: P^t
- Load on lower chord: P^b

Method of Joints and Method of Section Analysis are used to compute forces in truss elements without relying on the computer. Method of Joints is based on the equilibrium of the forces acting within the joint. Method of Section Analysis is based on the equilibrium of the forces acting from either the left or the right of the section. $\left(\sum x = 0,\ \sum y = 0,\ \sum M = 0\right)$.

The truss joints are assumed to be hinges, and the loads acting on the truss are represented as forces concentrated within the truss joints.

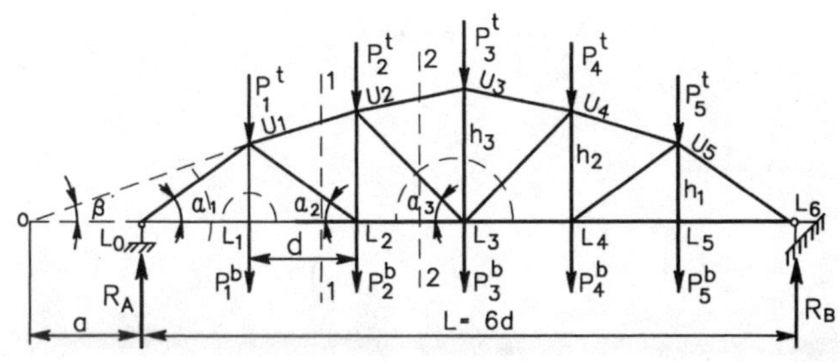

Member	Joints	Forces
L_0U_1 L_0L_1	**Joint L_0**	$R_A = \frac{d}{L}\Big[\left(P_1^t + P_1^b\right)\cdot 5 + \left(P_2^t + P_2^b\right)\cdot 4 + \left(P_3^t + P_3^b\right)\cdot 3 +$ $+\left(P_4^t + P_4^b\right)\cdot 2 + \left(P_5^t + P_5^b\right)\Big], \quad R_B = R_A - \sum_{n=1}^{n=5}\left(P_i^t + P_i^b\right).$ $\sum Y = R_A + L_0U_1 \cdot \sin\alpha_0 = 0,$ $L_0U_1 = -R_A / \sin\alpha_0$ (compression) . $\sum X = -L_0U_1 \cdot \cos\alpha_0 + L_0L_1 = 0,$ $L_0L_1 = L_0U_1 \cdot \cos\alpha_0$ (tension).
U_1L_1 L_1L_2	**Joint L_1**	$\sum Y = U_1L_1 - P_1^b = 0, \qquad U_1L_1 = P_1^b$ (tension). $\sum X = -L_0L_1 + L_1L_2 = 0, \quad L_1L_2 = L_0L_1$ (tension).
U_1L_2	**Section 1–1**	$\tan\beta = \frac{h_2 - h_1}{d}, \quad a = \frac{h_1}{\tan\beta} - d, \quad r_1 = (a + 2d)\sin\alpha_2 .$ $\sum M_O = U_1L_2r_1 - R_Aa + \left(P_1^t + P_1^b\right)(a + d) = 0,$ $U_1L_2 = \frac{1}{r_1}\left[R_Aa - \left(P_1^t + P_1^b\right)(a + d)\right]$ (compression or tension)

NOTES

TRUSSES

METHOD OF JOINTS and METHOD OF SECTION ANALYSIS

EXAMPLES

6.2

Member	Joints	Forces
U_1U_2	Section 1 – 1 (cont.) 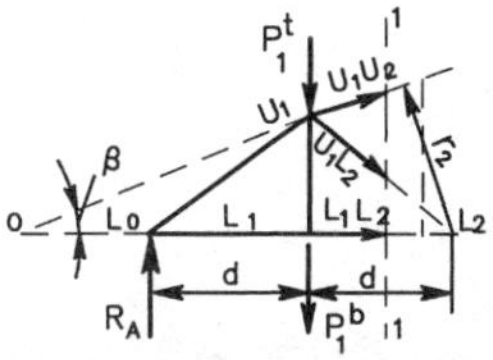	$r_2 = (a + 2d)\sin\beta$ $\sum M_{L2} = U_1U_2 r_2 + R_A 2d - (P_1^t + P_1^b)d = 0$, $U_1U_2 = -R_A 2d - (P_1^t + P_1^b)d$ (compression).
U_2L_2 L_2L_3	Joint L_2 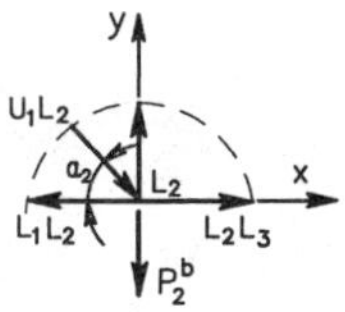	$\sum Y = U_2L_2 - U_1L_2 \sin\alpha_2 - P_2^b = 0$, $U_2L_2 = P_2^b + U_1L_2 \sin\alpha_2$ (tension). $\sum X = -L_1L_2 + L_2L_3 + U_1L_2 \cos\alpha_2 = 0$, $L_2L_3 = L_1L_2 - U_1L_2 \cos\alpha_2$ (tension).
U_2L_3 L_2L_3	Section 2 – 2 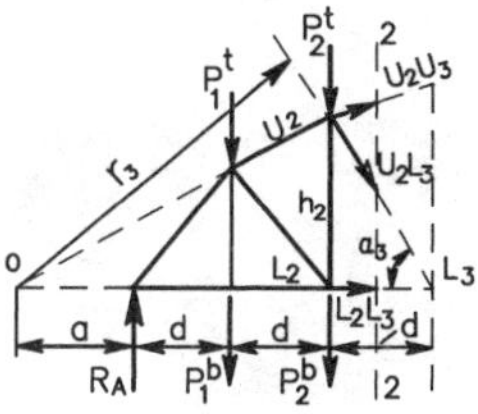	$r_3 = (a + 3d)\sin\alpha_3$ $\sum M_O = U_2L_3 r_3 - R_A a + (P_1^t + P_b^1)(a + d) +$ $+ (P_2^t + P_2^b)(a + 2d) = 0$, $U_2L_3 = \frac{1}{r_3}\left[R_A a - (P_1^t + P_1^b)(a + d) - (P_2^t + P_2^b)(a + 2d)\right]$ (compression). $\sum M_{U2} = -L_2L_3 h_2 + R_A 2d - (P_1^t + P_1^b)d = 0$, $L_2L_3 = \frac{1}{h_2}\left[R_A 2d - (P_1^t + P_1^b)d\right]$ (tension).
U_3L_3	Joint L_3 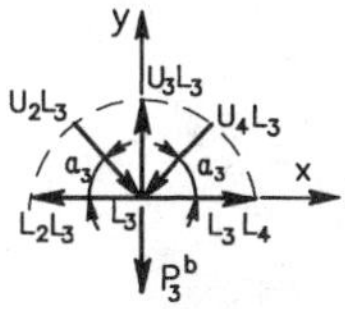	If $P_4^t = P_2^t,\ P_5^t = P_1^t,\ P_4^b = P_2^b,\ P_5^t = P_1^b$, $L_3L_4 = L_2L_3,\ \ U_4L_3 = U_2L_3$ $\sum Y = U_3L_3 - U_2L_3 \sin\alpha_3 - U_4L_3 \sin\alpha_3 - P_3^b = 0$ $U_3L_3 = P_3^b + U_2L_3 \sin\alpha_3 + U_4L_3 \sin\alpha_3$ (tension).

NOTES

INFLUENCE LINES (EXAMPLES)

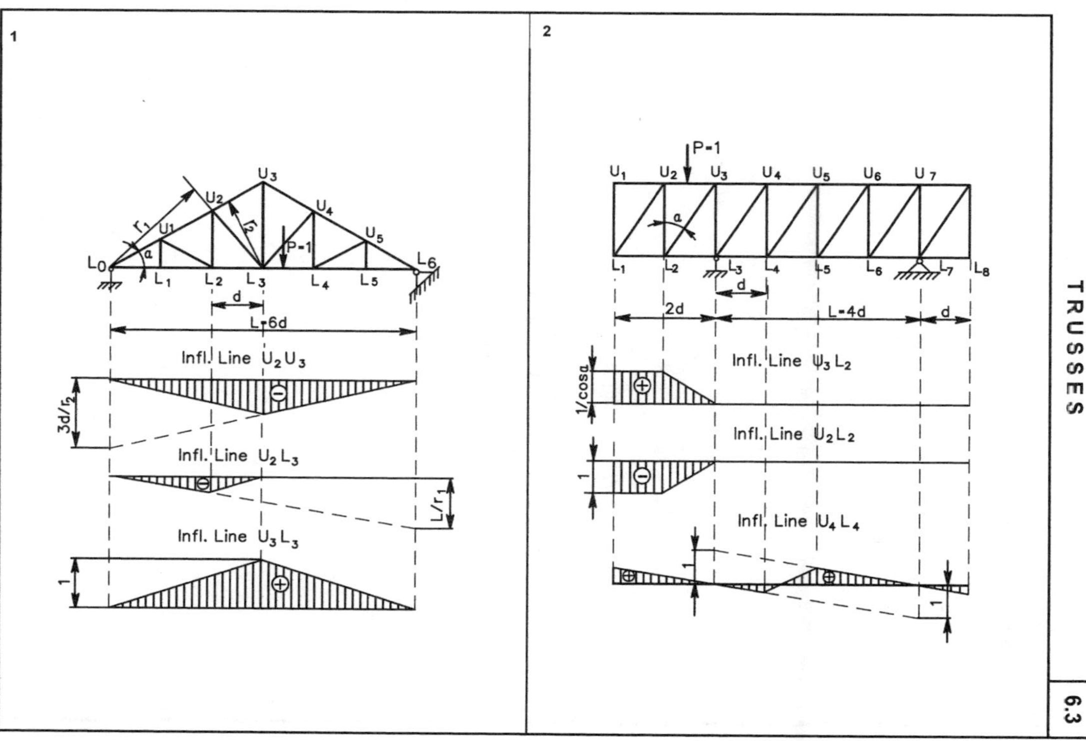

N O T E S

Example. Computation of truss

Given. Truss 3 in Table 6.4, $L = 12\ \text{m},\ \ d = 2\ \text{m},\ \ h = 4\ \text{m}$

$$\tan\alpha = \frac{h}{d} = 2.0,\quad \alpha = 63.435^0,\quad \cos\alpha = 0.447$$

Load: $P_2^b = P_6^b = 3\ \text{kN},\ \ P_3^b = P_5^b = 4\ \text{kN},\ \ P_4^b = 5\ \text{kN}$

Required. Compute forces in truss members using influence lines

Solution. $\frac{2d}{h} = \frac{2\times2}{4} = 1,\quad \frac{d}{h} = \frac{2}{4} = 0.5,\quad \frac{1}{\cos\alpha} = 2.237$

$$U_4U_5 = \frac{2d}{h}P_4^b + 2\frac{2d}{h(0.5L)}\times 2d\times P_3^b + 2\frac{2d}{h(0.5L)}\times d\times P_2^b$$

$$= 5 + 1.333\times4 + 0.667\times3 = 12.33\ \text{kN}\ \ \text{(compression)}$$

$$L_2L_3 = \frac{d}{hL}\times d\left(5\times P_2^b + 4\times P_3^b + 3\times P_4^b + 2\times P_5^b + P_6^b\right)$$

$$= 0.083\times(5\times3 + 4\times4 + 3\times5 + 2\times4 + 3) = 4.73\ \text{kN}\ \ \text{(tension)}$$

$$U_2L_3 = \frac{2.237}{L}d\left(-P_2^b + P_6^b + 2\times P_5^b + 3\times P_4^b + 4\times P_3^b\right)$$

$$= 0.3728(-3 + 3 + 2\times4 + 3\times5 + 4\times4) = 14.53\ \text{kN}\ \ \text{(tension)}$$

$$U_4L_3 = -U_2L_3 = -14.57\ \text{kN}\ \ \text{(compression)}$$

INFLUENCE LINES (EXAMPLES)

3

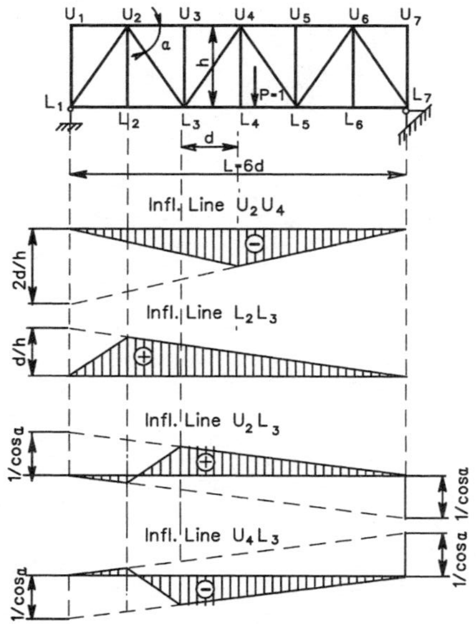

U_1 U_2 U_3 U_4 U_5 U_6 U_7
L_1 L_2 L_3 L_4 L_5 L_6 L_7
a h P=1 d L=6d

Infl. Line U_2U_4 — 2d/h

Infl. Line L_2L_3 — d/h

Infl. Line U_2L_3 — 1/cosa, 1/cosa

Infl. Line U_4L_3 — 1/cosa, 1/cosa

4

U_1 U_2 U_3 U_4 U_5 U_6 U_7
L_1 L_2 L_3 L_4 L_5 L_6 L_7
m a d L=6d

Infl. Line mL_4 — 1/cos

Infl. Line L_3m — 1/sin a, 1/sin a

Infl. Line U_2L_2 — 1

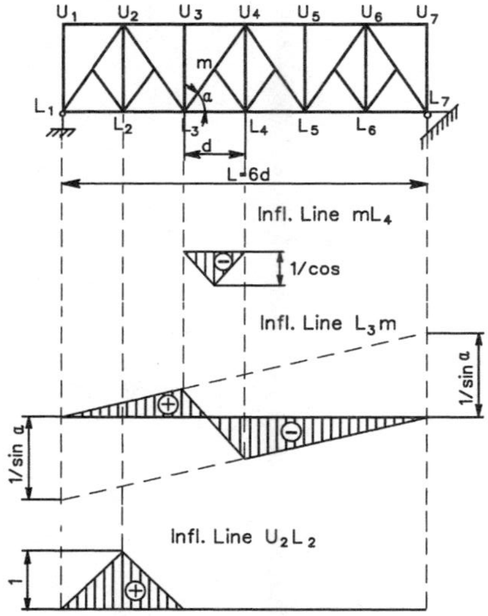

NOTES

7. PLATES

Bending Moments for Various Support and Loading Conditions

NOTES

Tables 7.1–7.9 provide formulas and coefficients for computation of bending moments in elastic plates.

The calculations are performed for plates of 1 meter width.

The plates are analyzed in two directions for various support conditions and acting loads.

Units of measurement: Distributed loads (w): kN/m^2

Bending moments (M): $kN \cdot m/m$

RECTANGULAR PLATES

BENDING MOMENTS

7.1

CASE A: $\frac{b}{a} > 2$ **CASE B:** $\frac{b}{a} \le 2$

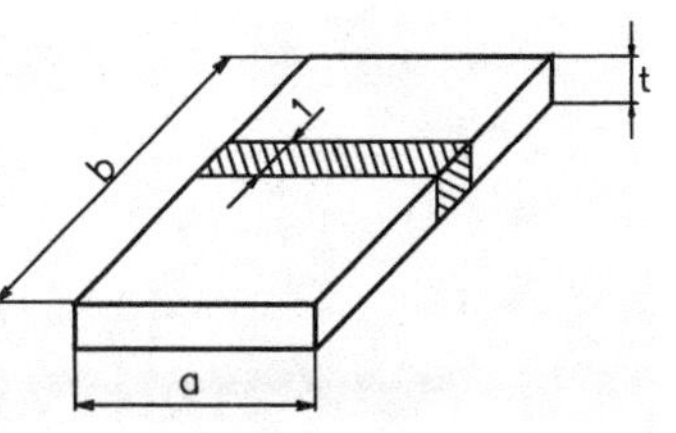

a < b

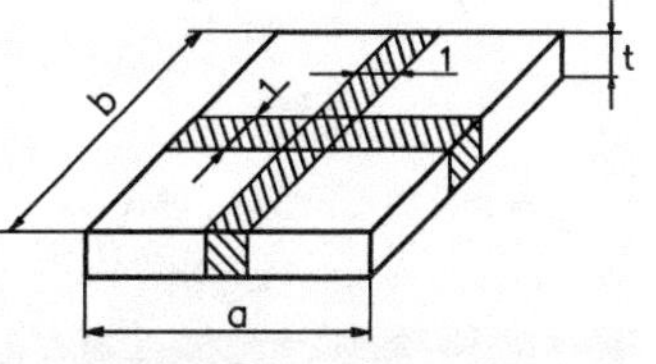

Case A $\frac{b}{a} > 2$ Plate should be computed in one (short) direction as a beam of length L = a

Case B $\frac{b}{a} \le 2$ Plate should be computed in two directions as two beams of lengths $L_1 = a$ and $L_2 = b$

Formulas for bending moments computation $\left(\frac{b}{a} \le 2\right)$

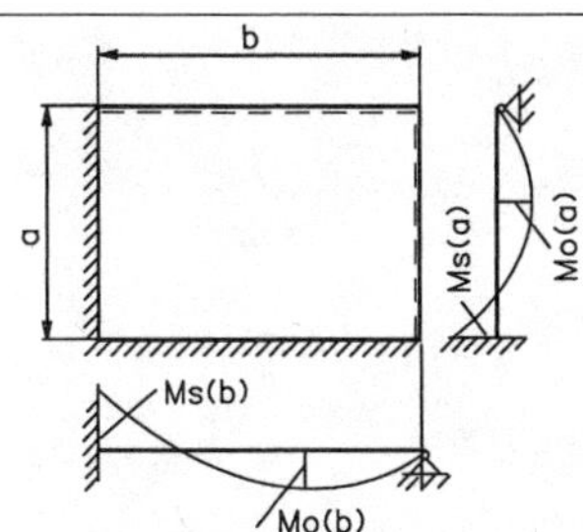

$$M_{0(a)} = \alpha_a \cdot w \cdot a \cdot b, \quad M_{0(b)} = \alpha_b \cdot w \cdot a \cdot b$$

$$M_{s(a)} = \beta_a \cdot w \cdot a \cdot b, \quad M_{s(b)} = \beta_b \cdot w \cdot a \cdot b$$

Where: w = uniformly distributed load

$\alpha_a, \alpha_b, \beta_a, \beta_b$ = coefficients from tables

for Poisson's ratio $\mu_T = 0$

Bending moments for any Poisson's ratio μ:

$$M^{\mu}_{(a)} = \frac{1}{1-\mu_T^2}\left[(1-\mu\mu_T)M_{(a)} + (\mu-\mu_T)M_{(b)}\right], \qquad M^{\mu}_{(b)} = \frac{1}{1-\mu_T^2}\left[(1-\mu\mu_T)M_{(b)} + (\mu-\mu_T)M_{(a)}\right]$$

Support condition

Legend:

(hatched line)	Plate fixed along edge.
(dashed line)	Plate hinged along edge.
(solid line)	Plate free along edge.
(column symbol)	Plate supported on column.

N O T E S

Example. Computation of rectangular plate, $b \leq 2a$

Given. Elastic steel plate 3 in Table 7.2, $a = 1.5\text{ m}$, $b = 2.1\text{ m}$, $t = 0.04\text{ m}$, $b/a = 1.4$

Uniformly distributed load $w = 0.8\text{ kN/m}^2$

Poisson's ratio $\mu = \mu_T = 0$

Required. Compute bending moments $M_{0(a)}$, $M_{0(b)}$, $M_{s(a)}$, $M_{s(b)}$

Solution.

$$M_{0(a)} = \alpha_a wab = 0.0323 \times 0.8 \times 1.5 \times 2.1 = 0.0814\text{ kN}\cdot\text{m/m} = 81.4\text{ N}\cdot\text{m/m}$$

$$M_{0(b)} = \alpha_b wab = 0.0165 \times 0.8 \times 1.5 \times 2.1 = 0.0416\text{ kN}\cdot\text{m/m} = 41.6\text{ N}\cdot\text{m/m}$$

$$M_{s(a)} = \beta_a wab = -0.0709 \times 0.8 \times 1.5 \times 2.1 = -0.1787\text{ kN}\cdot\text{m/m} = -178.7\text{ N}\cdot\text{m/m}$$

$$M_{s(b)} = \beta_b wab = -0.0361 \times 0.8 \times 1.5 \times 2.1 = -0.0910\text{ kN}\cdot\text{m/m} = -91.0\text{ N}\cdot\text{m/m}$$

RECTANGULAR PLATES

BENDING MOMENTS (uniformly distributed load)

7.2

Plate supports	b/a	α_a	α_b	β_a	β_b
1 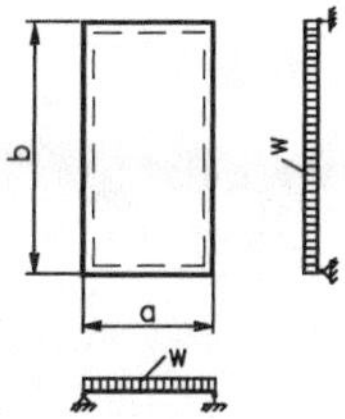	1.0	0.0363	0.0365		
	1.1	0.0399	0.0330		
	1.2	0.0428	0.0298		
	1.3	0.0452	0.0268		
	1.4	0.0469	0.0240		
	1.5	0.0480	0.0214		
	1.6	0.0485	0.0189		
	1.7	0.0488	0.0169		
	1.8	0.0485	0.0148		
	1.9	0.0480	0.0133		
	2.0	0.0473	0.0118		
2 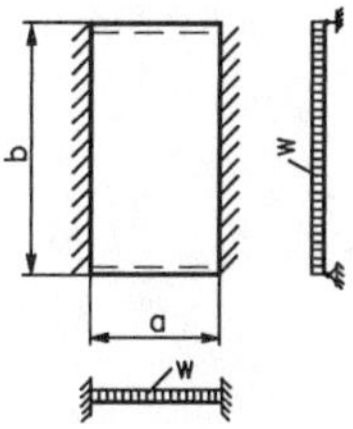	1.0	0.0267	0.0180	−0.0694	
	1.1	0.0266	0.0146	−0.0667	
	1.2	0.0261	0.0118	−0.0633	
	1.3	0.0254	0.0097	−0.0599	
	1.4	0.0245	0.0080	−0.0565	
	1.5	0.0235	0.0066	−0.0534	
	1.6	0.0226	0.0056	−0.0506	
	1.7	0.0217	0.0047	−0.0476	
	1.8	0.0208	0.0040	−0.0454	
	1.9	0.0199	0.0034	−0.0432	
	2.0	0.0193	0.0030	−0.0412	
3 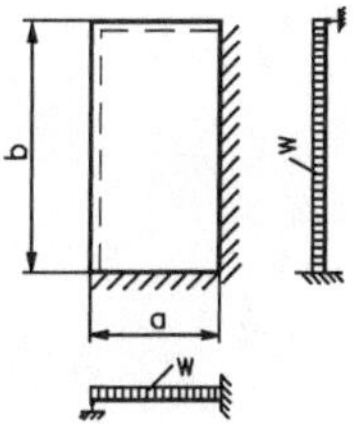	1.0	0.0269	0.0269	−0.0625	−0.0625
	1.1	0.0292	0.0242	−0.0675	−0.0558
	1.2	0.0309	0.0214	−0.0703	−0.0488
	1.3	0.0319	0.0188	−0.0711	−0.0421
	1.4	0.0323	0.0165	−0.0709	−0.0361
	1.5	0.0324	0.0144	−0.0695	−0.0310
	1.6	0.0321	0.0125	−0.0678	−0.0265
	1.7	0.0316	0.0109	−0.0657	−0.0228
	1.8	0.0308	0.0096	−0.0635	−0.0196
	1.9	0.0302	0.0084	−0.0612	−0.0169
	2.0	0.0294	0.0074	−0.0588	−0.0147

NOTES

RECTANGULAR PLATES

BENDING MOMENTS (uniformly distributed load) 7.3

Plate supports	b/a	α_a	α_b	β_a	β_b
4	1.0	0.0334	0.0273	−0.0892	
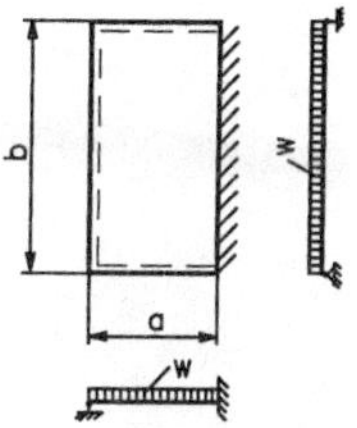	1.1	0.0349	0.0231	−0.0892	
	1.2	0.0357	0.0196	−0.0872	
	1.3	0.0359	0.0165	−0.0843	
	1.4	0.0357	0.0140	−0.0808	
	1.5	0.0350	0.0119	−0.0772	
	1.6	0.0341	0.101	−0.0735	
	1.7	0.0333	0.086	−0.0701	
	1.8	0.0326	0.0075	−0.0668	
	1.9	0.0316	0.0064	−0.0638	
	2.0	0.0303	0.0056	−0.0610	
5	1.0	0.0273	0.0334		−0.0893
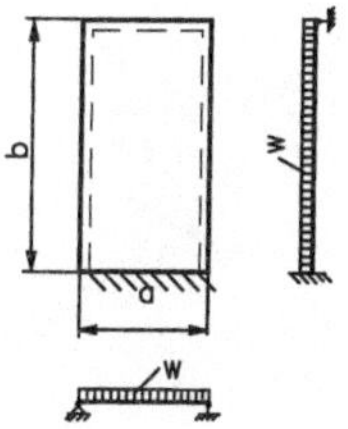	1.1	0.0313	0.0313		−0.0867
	1.2	0.0348	0.0292		−0.0820
	1.3	0.0378	0.0269		−0.0760
	1.4	0.0401	0.0248		−0.0688
	1.5	0.0420	0.0228		−0.0620
	1.6	0.0433	0.0208		−0.0553
	1.7	0.0441	0.0190		−0.0489
	1.8	0.0444	0.0172		−0.0432
	1.9	0.0445	0.0157		−0.0332
	2.0	0.0443	0.0142		−0.0338
	1.0	0.0226	0.0198	−0.0556	−0.0417
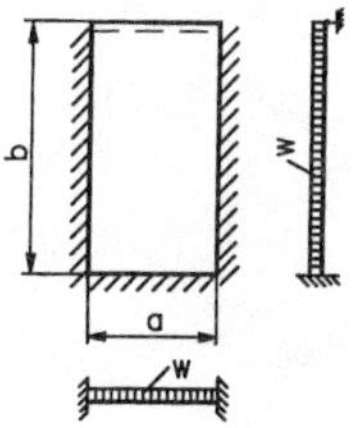	1.1	0.0234	0.0169	−0.0565	−0.0350
	1.2	0.0236	0.0142	−0.0560	−0.0292
	1.3	0.0235	0.0120	−0.0545	−0.0242
	1.4	0.0230	0.0102	−0.0526	−0.0202
	1.5	0.0225	0.0086	−0.0506	−0.0169
	1.6	0.0218	0.0073	−0.0484	−0.0142
	1.7	0.0210	0.0062	−0.0462	−0.0120
	1.8	0.0203	0.0054	−0.0442	−0.0102
	1.9	0.0192	0.0043	−0.0413	−0.0082
	2.0	0.0189	0.0040	−0.0404	−0.0076

NOTES

RECTANGULAR PLATES

BENDING MOMENTS (uniformly distributed load)

7.4

Plate supports	b/a	α_a	α_b	β_a	β_b
7 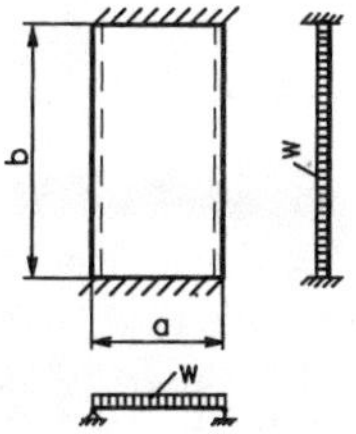	1.0	0.0180	0.0267		–0.0694
	1.1	0.0218	0.0262		–0.0708
	1.2	0.0254	0.0254		–0.0707
	1.3	0.0287	0.0242		–0.0689
	1.4	0.0316	0.0229		–0.0660
	1.5	0.0341	0.0214		–0.0621
	1.6	0.0362	0.0200		–0.0577
	1.7	0.0376	0.0186		–0.0531
	1.8	0.0388	0.0172		–0.0484
	1.9	0.0396	0.0158		–0.0439
	2.0	0.0400	0.0146		–0.0397
8 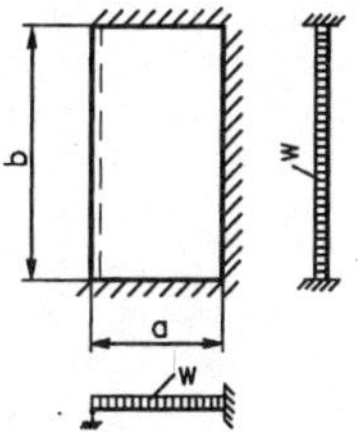	1.0	0.0198	0.0226	–0.0417	–0.0556
	1.1	0.0226	0.0212	–0.0481	–0.0530
	1.2	0.0249	0.0198	–0.0530	–0.0491
	1.3	0.0266	0.0181	–0.0565	–0.0447
	1.4	0.0279	0.0162	–0.0588	–0.0400
	1.5	0.0285	0.0146	–0.0597	–0.0354
	1.6	0.0289	0.0130	–0.0599	–0.0312
	1.7	0.0290	0.0116	–0.0594	–0.0274
	1.8	0.0288	0.0103	–0.0583	–0.0240
	1.9	0.0284	0.0092	–0.0570	–0.0212
	2.0	0.0280	0.0081	–0.0555	–0.0187
9 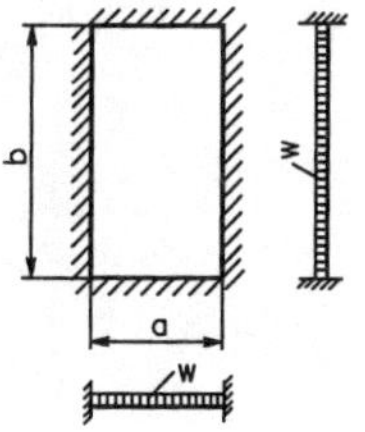	1.0	0.0179	0.0179	–0.0417	–0.0417
	1.1	0.0194	0.0161	–0.0450	–0.0372
	1.2	0.0204	0.0142	–0.0468	–0.0325
	1.3	0.0208	0.0123	–0.0475	–0.0281
	1.4	0.0210	0.0107	–0.0473	–0.0242
	1.5	0.0208	0.0093	–0.0464	–0.0206
	1.6	0.0205	0.0080	–0.0452	–0.0177
	1.7	0.0200	0.0069	–0.0438	–0.0152
	1.8	0.0195	0.0060	–0.0423	–0.0131
	1.9	0.0190	0.0052	–0.0408	–0.0113
	2.0	0.0183	0.0046	–0.0392	–0.0098

NOTES

RECTANGULAR PLATES

BENDING MOMENTS (uniformly distributed load)

7.5

Plate supports	b/a	α_a	α_b	β_a	β_b
10 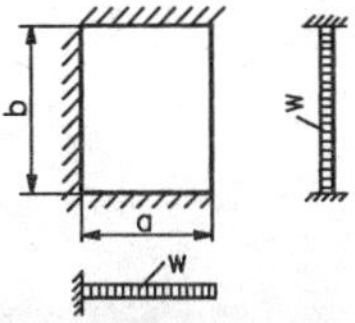	1.0	0.0099	0.0457	–0.0510	–0.0853
	1.1	0.0102	0.0492	–0.0574	–0.0930
	1.2	0.0102	0.0519	–0.0636	–0.1000
	1.3	0.0100	0.0540	–0.0700	–0.1062
	1.4	0.0097	0.00552	–0.0761	–0.1115
	1.5	0.0095	0.0556	–0.0821	–0.1155
11 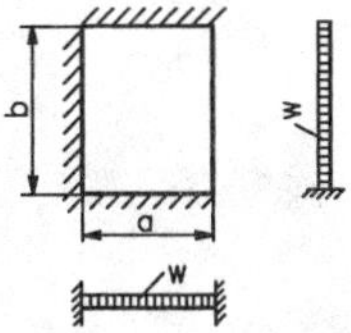	1.0	0.0457	0.0099	–0.0853	–0.0510
	1.1	0.0421	0.0094	–0.0777	–0.0448
	1.2	0.0389	0.0087	–0.0712	–0.0397
	1.3	0.0362	0.0079	–0.0658	–0.0354
	1.4	0.0362	0.0070	–0.0609	–0.0314
	1.5	0.0311	0.0059	–0.0562	–0.0279

BENDING MOMENTS (concentrated load at center)

$M_{0(a)} = \alpha_a \cdot P, \quad M_{0(b)} = \alpha_b \cdot P, \quad M_{s(a)} = \beta_a \cdot P, \quad M_{s(b)} = \beta_b \cdot P$

Plate supports	b/a	α_a	α_b	β_a	β_b
1 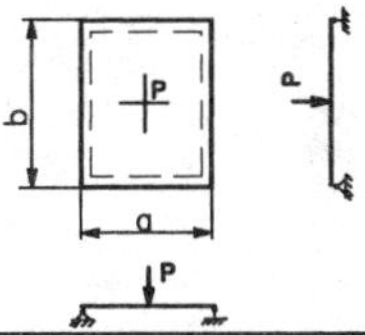	1.0	0.146	0.146		
	1.2	0.179	0.141		
	1.4	0.214	0.138		
	1.6	0.244	0.135		
	1.8	0.270	0.132		
	2.0	0.290	0.130		
2 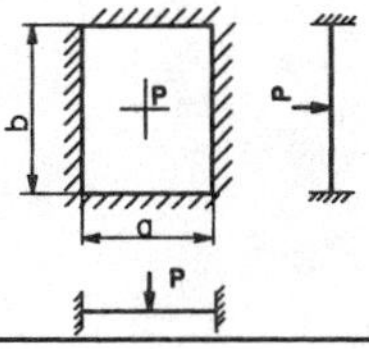	1.0	0.108	0.108	–0.094	–0.094
	1.2	0.128	0.100	–0.126	–0.074
	1.4	0.143	0.092	–0.149	–0.055
	1.6	0.156	0.086	–0.162	–0.040
	1.8	0.162	0.080	–0.171	–0.030
	2.0	0.168	0.076	–0.176	–0.022

N O T E S

Example. Computation of rectangular plate, $b \leq 2a$

Given. Elastic plate 1 in Table 7.6, $a = 1.8$ m, $b = 2.25$ m, $t = 0.1$ m, $a/b = 0.8$

Modulus of elasticity $E = 4030 \text{ kip/in}^2 = \dfrac{4030 \times 4.48222}{2.54^2} = 2800 \text{ kN/cm}^2$

Poisson's ratio $\mu = \mu_T = 1/6$,

Elastic stiffness $D = \dfrac{Et^3}{12(1-\mu^2)} = \dfrac{2800 \times 10^3}{12\left[1-(1/6)^2\right]} = 240000$

Uniformly distributed load $w = 0.2 \text{ kN/m}^2 = 0.002 \text{ kN/cm}^2$

Required. Compute bending moments $M_{0(a)}$ and $M_{0(b)}$, deflection Δ_0

Solution. $M_{0(a)} = \alpha_a w b^2 = 0.0323 \times 0.2 \times 2.25^2 = 0.0327 \text{ kN} \cdot \text{m/m} = 32.7 \text{ N} \cdot \text{m/m}$

$M_{0(b)} = \alpha_b w b^2 = 0.1078 \times 0.2 \times 2.25^2 = 0.1091 \text{ kN} \cdot \text{m/m} = 109.1 \text{ N} \cdot \text{m/m}$

$$\Delta_0 = \eta_0 w \frac{b^4}{D} = 0.018 \times 0.002 \times \frac{225^4}{24000} = 0.38 \text{ cm} = 3.8 \text{ mm}$$

RECTANGULAR PLATES

BENDING MOMENTS and DEFLECTIONS (uniformly distributed load) 7.6

$$M_{0(a)} = \alpha_a \cdot w \cdot b^2, \quad M_{0(b)} = \alpha_b \cdot w \cdot b^2, \quad M_{1(a)} = \alpha_{1(a)} \cdot w \cdot b^2, \quad M_{2(b)} = \alpha_{2(b)} \cdot w \cdot b^2$$

α_a, α_b, $\alpha_{1(a)}$ and $\alpha_{2(b)}$ = coefficients for Poisson's ratio $\mu_T = 1/6$

$$\Delta_0 = \eta_0 \cdot w \cdot \frac{b^4}{D}, \quad \Delta_1 = \eta_1 \cdot w \cdot \frac{b^4}{D}, \quad \Delta_2 = \eta_2 \cdot w \cdot \frac{b^4}{D}, \quad D = \frac{E \cdot t^3}{12(1-\mu^2)}$$

Where Δ_i = deflection at point i, E = Modulus of elasticity

t = plate thickness, μ = Poisson's ratio

D = Elastic stiffness

Plate supports	a/b	$\alpha_{0(a)}$	$\alpha_{0(b)}$	$\alpha_{1(a)}$	$\alpha_{2(b)}$	η_0	η_1	η_2
1 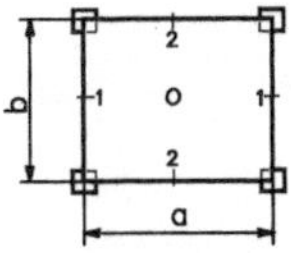	1.0	0.0947	0.0947	0.1606	0.1606	0.0263	0.0172	0.0172
	0.9	0.0689	0.1016	0.1367	0.1541	0.0218	0.0119	0.0164
	0.8	0.0479	0.1078	0.1148	0.1486	0.0180	0.0079	0.0157
	0.7	0.0289	0.1132	0.0955	0.1435	0.0158	0.0050	0.0151
	0.6	0.0131	0.1178	0.0769	0.1386	0.0148	0.0030	0.0146
	0.5	0.0005	0.1214	0.0592	0.1339	0.0140	0.0016	0.0141
2 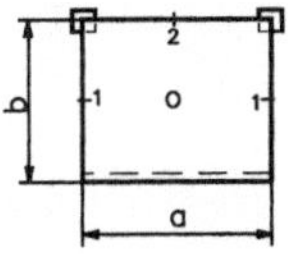	1.0	0.0977	0.1070	0.1578	0.2326	0.0606	0.0168	0.1011
	0.9	0.1007	0.0889	0.1552	0.2073	0.0418	0.0165	0.0625
	0.8	0.1038	0.0729	0.1526	0.1844	0.0307	0.0162	0.0406
	0.7	0.1069	0.0589	0.1498	0.1639	0.0247	0.0159	0.0275
	0.6	0.1097	0.0468	0.1470	0.1462	0.0209	0.155	0.0194
	0.5	0.1121	0.0364	0.1444	0.1314	0.185	0.0152	0.0142
3 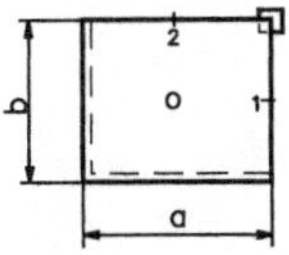	1.0	0.0581	0.0581	0.1198	0.1198	0.0122	0.0126	0.0126
	0.9	0.0500	0.0540	0.1031	0.1092	0.0100	0.0089	0.0117
	0.8	0.0421	0.0490	0.0866	0.0986	0.0080	0.0059	0.0106
	0.7	0.0343	0.0432	0.0706	0.0870	0.0063	0.0037	0.0093
	0.6	0.0270	0.0367	0.0547	0.0739	0.0048	0.0022	0.0078
	0.5	0.0202	0.0294	0.0388	0.0578	0.0036	0.0011	0.0063

NOTES

RECTANGULAR PLATES

BENDING MOMENTS (uniformly varying load)

7.7

$$M_{0(a)} = \alpha_a \cdot w \cdot \left(\frac{a \cdot b}{2}\right), \quad M_{0(b)} = \alpha_b \cdot w \cdot \left(\frac{a \cdot b}{2}\right), \quad M_{s(a)} = \beta_a \cdot w \cdot \left(\frac{a \cdot b}{2}\right), \quad M_{s(b)} = \beta_b \cdot w \cdot \left(\frac{a \cdot b}{2}\right)$$

Plate supports	b/a	α_a	α_b	β_a	β_b
1 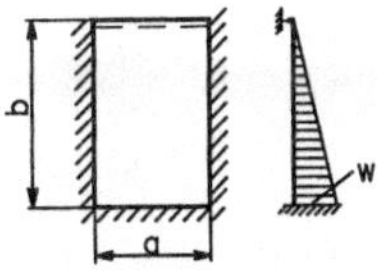	1.0	0.0216	0.0194	–0.0502	–0.0588
	1.1	0.0229	0.0178	–0.0515	–0.0554
	1.2	0.0236	0.0161	–0.0521	–0.0517
	1.3	0.0239	0.0145	–0.0522	–0.0477
	1.4	0.0241	0.0131	–0.0519	–0.0432
	1.5	0.0241	0.0117	–0.0514	–0.0387
2 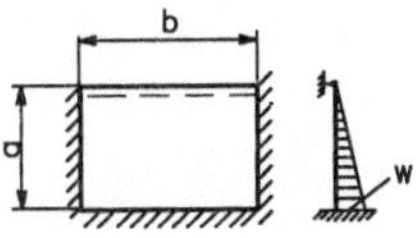	1.0	0.0194	0.0216	–0.0588	–0.0502
	1.1	0.0211	0.0198	–0.0614	–0.0480
	1.2	0.0228	0.0178	–0.0633	–0.0435
	1.3	0.0243	0.0153	–0.0644	–0.0418
	1.4	0.0257	0.0132	–0.0650	–0.0396
	1.5	0.0271	0.0120	–0.0652	–0.0357
3 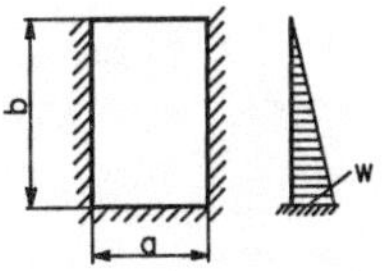	1.0	0.0246	0.0172	–0.0538	–0.0598
	1.1	0.0248	0.0163	–0.0538	–0.0553
	1.2	0.0250	0.0153	–0.0535	–0.0510
	1.3	0.0250	0.0142	–0.0529	–0.0469
	1.4	0.0247	0.0128	–0.0522	–0.0429
	1.5	0.0245	0.0114	–0.0514	–0.0390
4 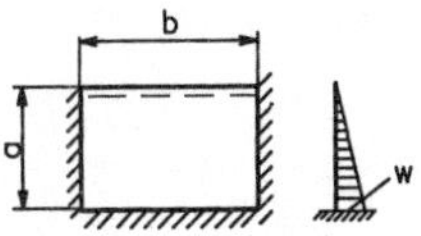	1.0	0.0172	0.0246	–0.0598	–0.0538
	1.1	0.0178	0.0244	–0.0640	–0.0535
	1.2	0.0180	0.0242	–0.0677	–0.0533
	1.3	0.0182	0.0244	–0.0709	–0.0533
	1.4	0.0180	0.0249	–0.0739	–0.0536
	1.5	0.0177	0.0262	–0.0765	–0.555

N O T E S

RECTANGULAR PLATES

BENDING MOMENTS (uniformly varying load)

7.8

$$M_{0(a)} = \alpha_a \cdot w \cdot \left(\frac{a \cdot b}{2}\right), \quad M_{0(b)} = \alpha_b \cdot w \cdot \left(\frac{a \cdot b}{2}\right), \quad M_{s(a)} = \beta_a \cdot w \cdot \left(\frac{a \cdot b}{2}\right), \quad M_{s(b)} = \beta_b \cdot w \cdot \left(\frac{a \cdot b}{2}\right)$$

Plate supports	b/a	α_a	α_b	β_a	β_b
5	1.0	0.0718	0.0042	–0.1412	–0.0422
	1.1	0.0672	0.0037	–0.1308	–0.0350
	1.2	0.0634	0.0031	–0.1222	–0.0290
	1.3	0.0598	0.0025	–0.1143	–0.0240
	1.4	0.0565	0.0019	–0.1069	–0.0200
	1.5	0.0530	0.0012	–0.1003	–0.0168
6	1.0	0.0042	0.0718	–0.0422	–0.1412
	1.1	0.0047	0.0758	–0.0509	–0.1510
	1.2	0.0053	0.0790	–0.0600	–0.1600
	1.3	0.0057	0.0810	–0.0692	–0.1675
	1.4	0.0060	0.0826	–0.0785	–0.1740
	1.5	0.0063	0.0828	–0.0876	–0.1790

$$M_{0(a)} = \alpha_a \cdot w \cdot \left(\frac{a \cdot b}{2}\right), \quad M_{0(b)} = \alpha_b \cdot w \cdot \left(\frac{a \cdot b}{2}\right)$$

$$M_{s(1)} = \beta_1 \cdot w \cdot \left(\frac{a \cdot b}{2}\right), \quad M_{s(2)} = \beta_2 \cdot w \cdot \left(\frac{a \cdot b}{2}\right), \quad M_{s(3)} = \beta_3 \cdot w \cdot \left(\frac{a \cdot b}{2}\right)$$

Plate supports	b/a	α_a	α_b	β_1	β_2	β_3
7	1.0	0.0184	0.0206	–0.0448	–0.0562	–0.0332
	1.1	0.0205	0.0190	–0.0477	–0.0538	–0.0302
	1.2	0.0221	0.0173	–0.0495	–0.0506	–0.0271
	1.3	0.0229	0.0156	–0.0504	–0.0470	–0.0237
	1.4	0.0235	0.0137	–0.0508	–0.0431	–0.0204
	1.5	0.0241	0.0120	–0.0510	–0.0387	–0.0168
8	1.0	0.0206	0.0184	–0.0562	–0.0332	–0.0446
	1.1	0.0218	0.0160	–0.0576	–0.0353	–0.0411
	1.2	0.0227	0.0137	–0.0580	–0.0357	–0.0372
	1.3	0.0231	0.0112	–0.0577	–0.0376	–0.0336
	1.4	0.0233	0.0090	–0.0569	–0.0380	–0.0302
	1.5	0.0233	0.0072	–0.0556	–0.0382	–0.0276

NOTES

CIRCULAR PLATES

BENDING MOMENTS, SHEAR and DEFLECTION (uniformly distributed load) — 7.9

a = circular plate's radius

r = circular section's radius

t = thickness of plate

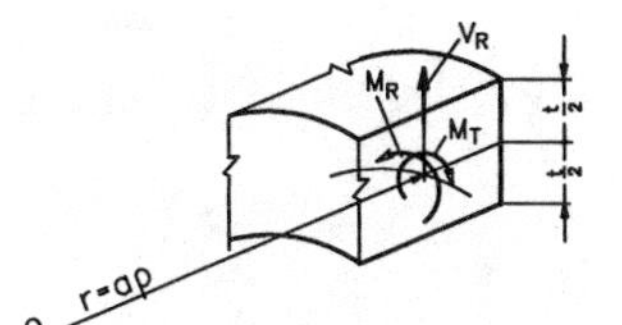

M_R = radial moment

M_T = tangential moment

V_R = radial shear

R = support reaction

Δ = deflection at center of plate

μ = Poisson's ratio

E = modulus of elasticity

Moment, shear and deflection diagrams	Formulas

1

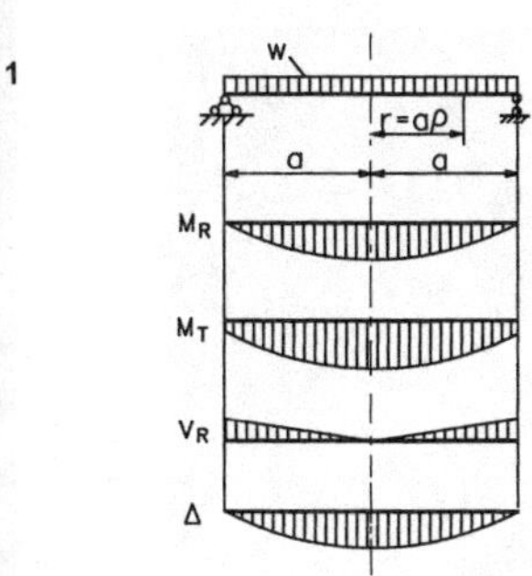

$$\rho = \frac{a}{r}, \quad P = w\pi a^2, \quad R = \frac{P}{2\pi a}, \quad V_R = -\frac{P}{2\pi a}\rho$$

$$M_R = \frac{P}{16\pi}(3+\mu)(1-\rho^2)$$

$$M_T = \frac{P}{16\pi}\left[3+\mu-(1+3\mu)\rho^2\right]$$

$$\Delta = \frac{Pa^2}{64\pi D}(1-\rho^2)\left(\frac{5+\mu}{1+\mu}-\rho^2\right), \quad D = \frac{Et^3}{12(1-\mu^2)}$$

2

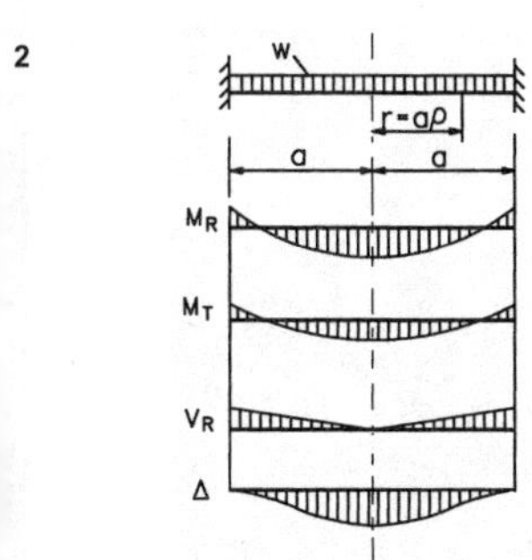

$$\rho = \frac{a}{r}, \quad P = w\pi a^2, \quad R = \frac{P}{2\pi a}, \quad V_R = -\frac{P}{2\pi a}\rho$$

$$M_R = \frac{P}{16\pi}\left[1+\mu-(3+\mu)\rho^2\right]$$

$$M_T = \frac{P}{16\pi}\left[1+\mu-(1+3\mu)\rho^2\right]$$

$$\Delta = \frac{Pa^2}{64\pi D}(1-\rho^2), \qquad D = \frac{Et^3}{12(1-\mu^2)}$$

NOTES

8. SOILS

N O T E S

For purposes of structural design, engineering properties of soils are determined through laboratory experiments and field research, conducted for specific conditions. If these methods are unavailable, use of data provided in the norms may be acceptable.

The modulus of deformation and Poisson's ratio of soil can be determined using the following formulas:

$$E_s = \frac{3c_1c_2}{2c_1 + c_2}, \qquad \mu = \frac{c_1 - c_2}{2c_1 + c_2}$$

$$c_1 = \frac{(1+2k_0)(1+e)}{D_r}, \qquad c_2 = \frac{(1-k_0)(1+e)}{D_r}$$

Where: k_0 = coefficient of lateral earth pressure (Table 10.1)
e = void ratio (Table 8.2)
D_r = relative density (Table 8.2)

Soil properties found in Tables 8.2–8.7 are provided only as guidelines.

SOILS

ENGINEERING PROPERTIES OF SOILS

8.1

SOIL TYPE	SOIL PARTICLES	
	SIZE	WEIGHT IN DRY SOIL
Cohesive soils Igneous and sedimentary stone compact soils; compact, sticky and plastic clay soils.	Less than 0.005 mm	
Cohesionless soils Crashed stone	Coarser than 10 mm	> 50 %
Gravel sand	Coarser than 2 mm	> 50 %
Coarse-grained sand	Coarser than 0.5 mm	> 50 %
Medium-grained sand	Coarser than 0.25 mm	> 50 %
Fine-grained sand	Coarser than 0.1 mm	> 75 %
Dustlike sand	Coarser than 0.1 mm	< 75 %

COMPONENTS OF SOIL

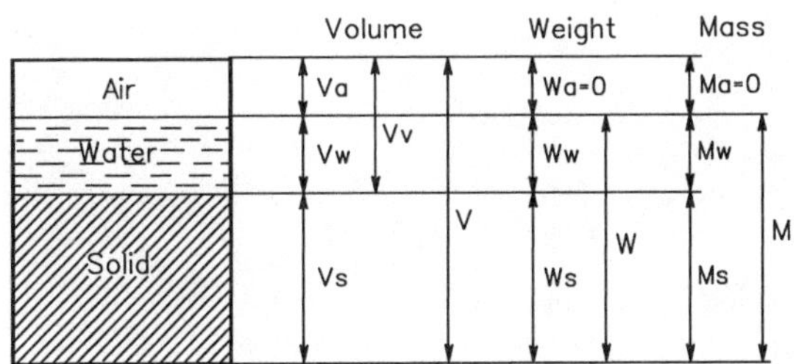

V, V_a, V_w, V_s and V_v = total volume and volume of air, water, solid matter and voids, respectively.

W, W_w and W_s = total weight and weight of water and solid matter, respectively.

M, M_w and M_s = total mass and mass of water and solid matter, respectively.

NOTES

SOILS

WEIGHT / MASS and VOLUME RELATIONSHIPS

8.2

1. Porosity: $n = \frac{V_v}{V} \cdot 100\ \%$, $V = V_s + V_v$	**9. Specific gravity of solids:**
2. Void ratio: $e = \frac{V_v}{V_s} = \frac{n}{1-n}$, $V_v = V_a + V_w$	$G_s = \frac{W_s / V_s}{\gamma_w} = \frac{W_s}{V_s \cdot \gamma_w}$ or $G_s = \frac{M_s / V_s}{\rho_w} = \frac{M_s}{V_s \cdot \rho_w}$ Where:
3. Degree of saturation: $S = \frac{V_w}{V_v} \cdot 100\ \%$	γ_w and ρ_w = unit weight and unit mass of water
4. Water content: $w = \frac{W_w}{W_s} \cdot 100\ \% = \frac{M_w}{M_s} \cdot 100\ \%$	$\gamma_w = 62.4\ \text{lb/ft}^3$ or $9.81\ \text{kN/m}^3$, $\rho_w = 1000\ \text{kg/m}^3$ (at normal temperatures)
5. Unit weight: $\gamma = \frac{W_s + W_w}{V}$	**10. Relative density:** $D_r = \frac{e_{max} - e_0}{e_{max} - e_{min}} \cdot 100\ \%$,
6. Dry unit weight: $\gamma_d = \frac{W_s}{V} = \frac{\gamma}{1+w}$	or $D_r = \frac{\gamma_{max}(\gamma - \gamma_{min})}{\gamma(\gamma_{max} - \gamma_{min})} \cdot 100\ \%$
7. Unit mass: $\rho = \frac{M}{V}$	Where: e_{max}, e_{min} and e_0 = maximum, minimum and in-place void ratio of the soil, respectively.
8. Dry unit mass: $\rho_d = \frac{M_s}{V}$	$\gamma_{max}, \gamma_{min}$ and γ_0 = maximum, minimum and in-place dry unit weight, respectively.

FLOW OF WATER IN SOIL

Darcy's Law.

Velocity of flow: $\upsilon = k_p \cdot i$,

where: k_p = coefficient of permeability,

$$i = \frac{\Delta H}{\Delta L} = \text{hydraulic gradient (slope)}.$$

Actual velocity:

$$\upsilon_{actual} = \frac{\upsilon}{n} = \frac{k_p \cdot i}{n} \quad \text{or} \quad \upsilon_{actual} = \frac{k_p \cdot i(1+e)}{e}.$$

Where:
n and e = soil's porosity and void ratio, respectively.

Flow rate (volume per unit time): $q = k_p \cdot i \cdot A$.

Where: A = area of the given cross-section of soil

COEFFICIENT OF PERMEABILITY (k_p)

SOIL TYPE	k_p cm / sec
Crashed stone, gravel sand	$1 \cdot 10^{-1}$
Coarse-grained sand	$1 \cdot 10^{-2}$ to $1 \cdot 10^{-1}$
Medium-grained sand	$1 \cdot 10^{-3}$ to $1 \cdot 10^{-2}$
Fine-grained sand	$1 \cdot 10^{-4}$ to $1 \cdot 10^{-3}$
Sandy loam	$1 \cdot 10^{-5}$ to $1 \cdot 10^{-3}$
Sandy clay	$1 \cdot 10^{-7}$ to $1 \cdot 10^{-5}$
Clay	$< 10^{-7}$

NOTES

STRESS DISTRIBUTION IN SOIL

Method based on elastic theory

Concentrated load

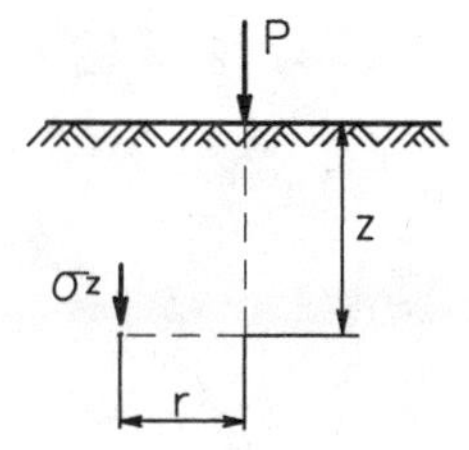

Boussinesq equation:

$$\sigma_z = \frac{3P}{2\pi z^2\left[1+(r/z)^2\right]^{5/2}},$$

Where σ_z = vertical stress at depth z

P = concentrated load

Uniformly distributed load

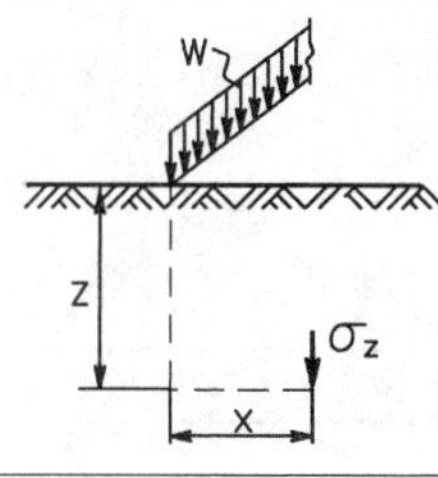

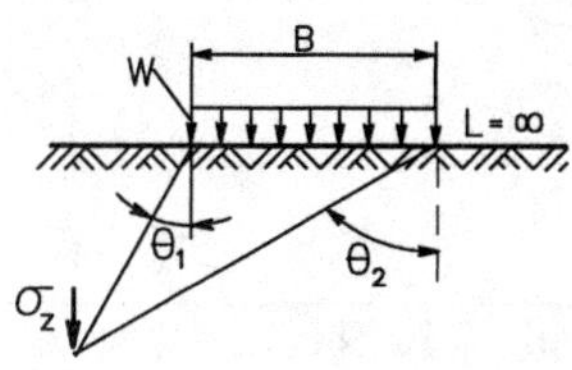

$$\sigma_z = \frac{2w}{\pi z\left[1+(x/z)^2\right]^2}$$

$$\sigma_z = \frac{w}{\pi}\left(\theta_2 - \theta_1 + \sin\theta_2\cos\theta_2 - \sin\theta_1\cos\theta_1\right)$$

Approximate method

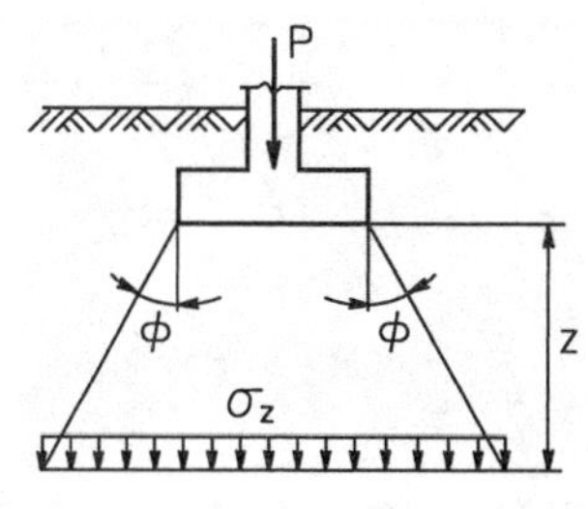

$$\sigma_z = \frac{P}{(B+2z\tan\phi)(L+2z\tan\phi)},$$

Where σ_z = approximate vertical stress at depth z

P = total load

B = width of footing

L = length of footing, B < L

z = depth

ϕ = angle of internal friction

N O T E S

Table 8.4

Example. Settlement of soil. Method based on elastic theory.

Units: $B(m)$, $L(m)$, $H(m)$, $h_i(m)$, $P_v(kN)$, $\gamma_i(kN/m^3)$, $\sigma_{a_i}(kPa)$, $E_{s_i}(kPa)$

P_v = weight of structures + weight of footing and surcharge + temporary load (live load)

z_i = distance from footing base to the middle of h_i layer

Lower border of active soil zone for vertical load P_v has been adopted as 20% of natural soil pressure: $0.2\sigma_\gamma$

Given. $B=3(m),\ L=5.4(m),\ H_1=5(m),\ h_0=2(m),\ h_1=h_2=h_3=1.0(m)<0.4B$

$H_2=4.0(m),\ h_4=h_5=h_6=h_7=1.0(m)<0.4B$

$\gamma_0=\gamma_1=1.8(ton/m^3)=17.7(kN/m^3),\ E_{s_1}=40000(kPa),\ \beta_1=0.76$

$\gamma_2=2.0(ton/m^3),\ E_{s_2}=25000(kPa),\ \beta_2=0.72$

Engineering properties of soils are determined by field and laboratory methods

Required. Compute settlement of soil under footing

Solution. $\sigma_p=\dfrac{P_v}{B\cdot L}=\dfrac{3000}{3\times5.4}=185.2(kPa),\ \sigma_{\gamma_0}=\gamma_0 h_0=17.7\times2.0=35.4(kPa)$

$\sigma_{a_0}=\sigma_p-\sigma_{\gamma_0}=185.2-35.4=149.8(kPa),\quad 0.2\sigma_\gamma=0.2\times\gamma_{1(2)}(h_0+z_i)\ (kPa)$

$\sigma_{a_i}=\alpha_i\times\sigma_{a_0}$, (for α_i see Table 8.5a), $\quad L/B=5.4/3.0=1.8$

H_i	$z_i(m)$	z_i/B	α_i	$\sigma_{a_i}(kPa)$	$0.2\sigma_\gamma(kPa)$
H_1	$z_1=0.5$	0.167	0.944	141.4	8.9
	$z_2=1.5$	0.500	0.794	118.9	12.4
	$z_3=2.5$	0.833	0.561	84.0	15.9
H_2	$z_4=3.5$	1.167	0.391	58.4	21.6
	$z_5=4.5$	1.500	0.282	42.2	25.5
	$z_6=5.5$	1.833	0.207	31.0	29.6
	$z_7=6.5$	2.167	0.157	23.5	33.3

Assume: $z=6.0(m),\ z/B=2.0,\ \alpha=0.189,$

$$\sigma_a=0.189\times149.8=28.3\approx0.2\sigma_\gamma=0.2(5.0\times17.7+3.0\times19.6)=29.5(kPa)$$

Settlement:

$$S=1.0(141.4+118.8+84.0)\frac{0.76}{40000}+1.0(58.4+42.2+31.0)\frac{0.72}{25000}=0.0065+0.0038=0.0103(m)$$

SETTLEMENT OF SOIL

Method based on elastic theory

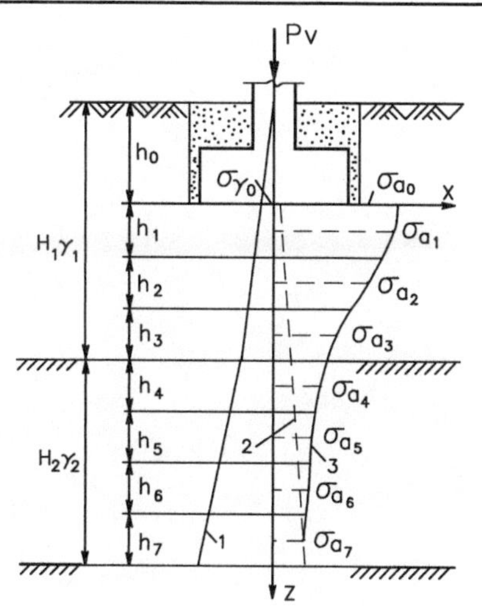

$1 = \text{Line } \sigma_\gamma$, $2 = \text{Line } 0.2\sigma_\gamma$, $3 = \text{Line } \sigma_a$

Settlement: $$S = \sum_{i=1}^{i=n} \sigma_{a_i} h_i \frac{\beta_i}{E_{s_i}}$$

Where

n = number of h-height layers, $h \le 0.4B$

σ_{a_i} = additional vertical pressure at the mid-height of h_i - layer, $\sigma_{a_i} = \alpha_i \cdot \sigma_{a_0}$

$$\sigma_{a_0} = \sigma_P - \sigma_{\gamma_0}, \quad \sigma_{\gamma_0} = \gamma_0 h_0, \quad \sigma_P = \frac{P_v}{B \cdot L}$$

α_i = coefficient from Table 8.5a

γ_i = unit weight of soil

P_v = total vertical load, $B < L$

B = width of footing, L = length of footing

E_{S_i} = modulus of deformation of soil

$$\beta = 1 - \frac{2\mu^2}{1-\mu}, \quad \mu = \text{Poisson's ratio for soil}$$

Sand: $\beta = 0.76$, Sandy loam: $\beta = 0.72$

Sandy clay: $\beta = 0.57$, Clay: $\beta = 0.4$

Alternative formulas

Settlement of loads on clay due to primary consolidation:

$$S = \frac{e_0 - e}{1 + e_0}[H]$$

e_0 = initial void ratio of the soil in situ

e = void ratio of the soil corresponding to the total pressure acting at midheight of the consolidating clay layer

H = thickness of the consolidating clay layer

Settlement of loads on clay due to secondary consolidation:

$$S_s = C_\alpha H \cdot \log(t_s / t_p), \quad C_\alpha \approx 0.01 - 0.03$$

t_s = life of the structure or time for which settlement is required

t_p = time to completion of primary consolidation

NOTES

Coefficient α_i Table 8.5a

z_i / B	L / B												For circle
	1.0	1.2	1.4	1.6	1.8	2.0	2.4	2.8	3.2	4.0	5.0	≥ 10	
0	1.000	1.000	1.000	1.000	1.000	1.000	1.000	1.000	1.000	1.000	1.000	1.000	1.000
0.4	0.800	0.830	0.848	0.859	0.866	0.870	0.875	0.878	0.879	0.880	0.881	0.881	0.756
0.8	0.449	0.496	0.532	0.558	0.578	0.593	0.612	0.623	0.630	0.636	0.639	0.642	0.390
1.2	0.257	0.294	0.325	0.352	0.374	0.392	0.419	0.437	0.469	0.462	0.470	0.477	0.214
1.6	0.160	0.187	0.210	0.232	0.251	0.267	0.294	0.314	0.329	0.348	0.360	0.374	0.130
2.0	0.108	0.127	0.145	0.161	0.176	0.189	0.214	0.233	0.241	0.270	0.285	0.304	0.087
2.4	0.077	0.092	0.105	0.118	0.130	0.141	0.161	0.178	0.192	0.213	0.230	0.258	0.062
2.8	0.058	0.069	0.079	0.089	0.099	0.108	0.124	0.139	0.152	0.172	0.189	0.228	0.046
3.2	0.045	0.053	0.062	0.070	0.077	0.085	0.098	0.110	0.122	0.141	0.158	0.190	0.036
3.6	0.036	0.042	0.049	0.056	0.062	0.068	0.080	0.090	0.100	0.117	0.133	0.175	0.030
4.0	0.029	0.035	0.040	0.046	0.051	0.056	0.066	0.075	0.084	0.095	0.113	0.158	0.025
4.4	0.024	0.029	0.034	0.038	0.042	0.047	0.055	0.063	0.070	0.084	0.098	0.144	0.021
4.8	0.020	0.024	0.028	0.032	0.036	0.040	0.047	0.054	0.060	0.072	0.085	0.132	0.018
5.0	0.019	0.022	0.026	0.030	0.033	0.037	0.044	0.050	0.056	0.067	0.079	0.126	0.017

Method based on Winkler's hypothesis

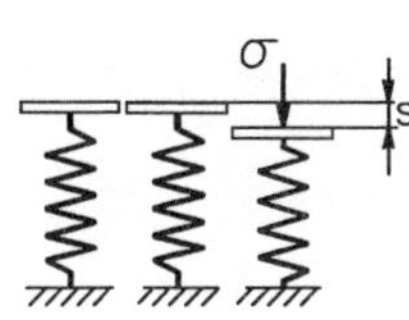

Winkler's support model

Settlement: $S = \dfrac{\sigma}{k_w}$

Where

σ = compressive stress applied to a unit area of a soil subgrade

S = settlement of unit area of a soil subgrade

k_w = Winkler's coefficient of subgrade reaction (force per length cubed)

$$k_w = \frac{E_s}{h}$$

Kw h, Es

$$k_{w_{1,2}} = \frac{k_{w_1} \cdot k_{w_2}}{k_{w_1} + k_{w_2}}$$

$$k_{w_1} = \frac{E_{s_1}}{h_1}, \qquad k_{w_2} = \frac{E_{s_2}}{h_2}$$

Kw_1 h_1, Es_1

Kw_2 h_2, Es_2

N O T E S

For slope stability analysis, it is necessary to compute the factor of safety for 2 or 3 possible failure surfaces with different diameters.

The smallest of the obtained values is then accepted as the result.

Modulus of deformation (E_s) and Winkler's coefficient (k_w) for some types of soil

Soil type	Range E_s (MPa)	Range k_w (N/cm^3)
Crashed stone, gravel sand	55 – 65	90 – 150
Coarse-grained sand	40 – 45	75 – 120
Medium-grained sand	35 – 40	60 – 90
Fine-grained sand	25 – 35	45 – 75
Sandy loam	15 – 25	30 – 60
Sandy clay	10 – 30	30 – 45
Clay	15 – 30	25 – 45

SHEAR STRENGTH OF SOIL

Coulomb equation: $\tau_s = c + \sigma \tan\phi$

Where τ_s = shear strength

c = cohesion

σ = effective intergranular normal pressure

ϕ = angle of internal friction

$\tan\phi$ = coefficient of friction

SLOPE STABILITY ANALYSIS

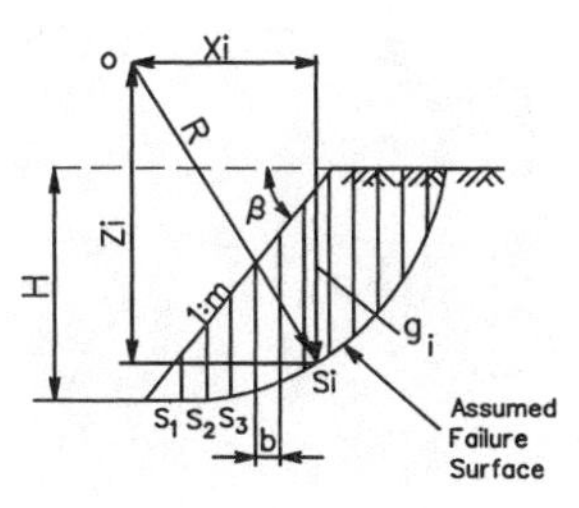

Factor of safety for slope F.S. ≥ 1.5 to 1.8

$$F.S. = \frac{\sum_{i=1}^{i=n} g_i z_i \tan\phi_i + R\sum_{i=1}^{i=n} c_i s_i}{\sum_{i=1}^{i=n} g_i x_i}$$

Where g_i = weight of mass for element i

c_i = cohesion of soil

ϕ_i = angle of internal friction

H = depth of cut

Safety depth of cut $H_s = \frac{2c}{\gamma}\cdot\frac{\cos\phi}{1-\sin\phi}$

N O T E S

Table 8.4

Example. Bearing capacity analysis

Given. Rectangular footing, $B = 3.6 \text{ m}, \quad L = 2.8 \text{ m}, \quad B/L = 1.28, \quad \text{smooth base}$

Granular soil, $\phi = 30^0, \quad c = 0, \quad \gamma = 130 \text{ Lb/ft}^3 = 130 \times 0.1571 = 20.42 \text{ kN/m}^3$

Loads $P = 2500 \text{ kN}, \quad M = 500 \text{ kN} \cdot \text{m}, \quad e = 500 / 2500 = 0.2 \text{ m}, \quad e/B = 0.2 / 3.6 = 0.06$

Bearing capacity factors $R_e = 0.78, \quad N_q = 20.1, \quad N_\gamma = 20$

Required. Compute factor of safety for footing

Solution. $q_{ult} = \gamma D_f N_q + 0.4 \gamma B N_\gamma = 20.42 \times 2 \times 20.1 + 0.4 \times 20.42 \times 3.6 \times 20 = 1409 \text{ kN/m}^2$

$F.S. = q_{ult} \cdot B \cdot L \cdot R_e / P = 1409 \times 3.6 \times 2.8 \times 0.78 / 2500 = 4.43 > 3$

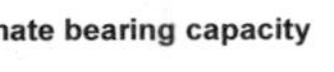

Eccentric load reduction factors Re

Ultimate bearing capacity

Continuous footing (width B):

$$q_{ult} = cN_c + \gamma D_f N_q + 0.5\gamma B N_\gamma$$

Square and rectangular footing (width B, length L):

$$q_{ult} = cN_c\left(1 + 0.3\frac{B}{L}\right) + \gamma D_f N_q + 0.4\gamma B N_\gamma$$

Circular footing (radius R):

$$q_{ult} = 1.3cN_c + \gamma D_f N_q + 0.6\gamma B N_\gamma$$

Where:

c = cohesion of soil

γ = unit weight of soil

N_c, N_q, N_γ = Terzaghi's bearing capacity factors

D_f = distance from ground surface to base of footing

Factor of safety for footing $F.S. \geq 2.5$ to 3

Continuous footing: $F.S. = q_{ult} \cdot B \cdot R_e / P$

Square and rectangular footing:

$$F.S. = q_{ult} \cdot B \cdot L \cdot R_e / P$$

Circular footing: $F.S. = q_{ult} \cdot \pi \cdot R^2 \cdot R_e / P$

Where R_e = eccentric load reduction factor

8.7

NOTES

9. FOUNDATIONS

NOTES

Tables 9.1–9.7 consider two cases of foundation analysis.

I. The footing is supported directly by the soil:

Maximum soil reaction (contact pressure) is determined and compared with requirements of the norms or the results of laboratory or field soil research.

II. The footing is supported by the piles:

Forces acting on the piles are computed and compared with the pile capacity provided in the catalogs.

If necessary, pile capacity can be computed using the formulas provided in Table 9.4.

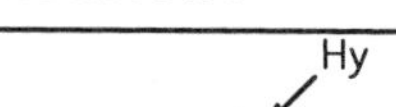

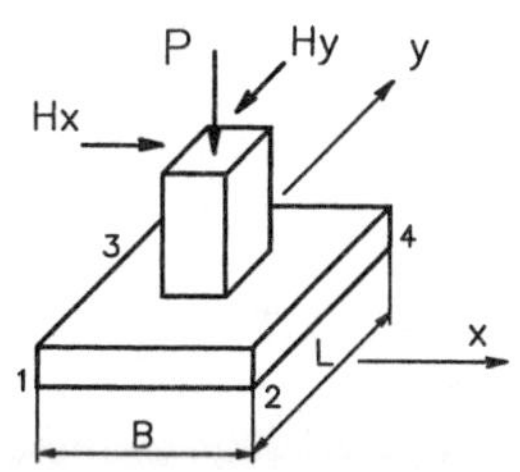

Individual column footing

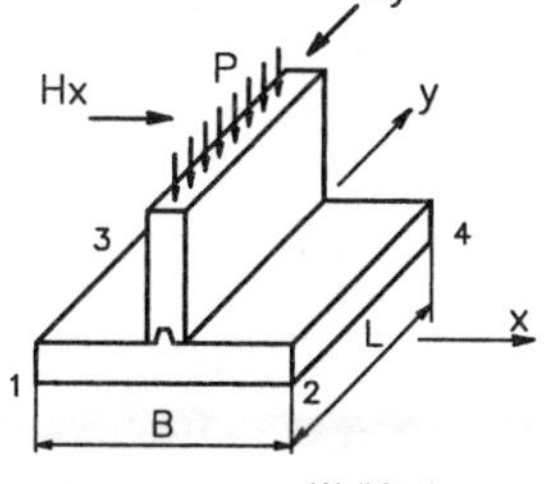

Wall footing

Contact pressure and soil pressure diagrams

Two-way action: $q_i = \dfrac{P_v}{A} \pm \dfrac{M_x}{S_x} \pm \dfrac{M_y}{S_y}$. Where $A = B \cdot L$, $S_x = \dfrac{B \cdot L^2}{6}$, $S_y = \dfrac{B^2 \cdot L}{6}$.

One-way action

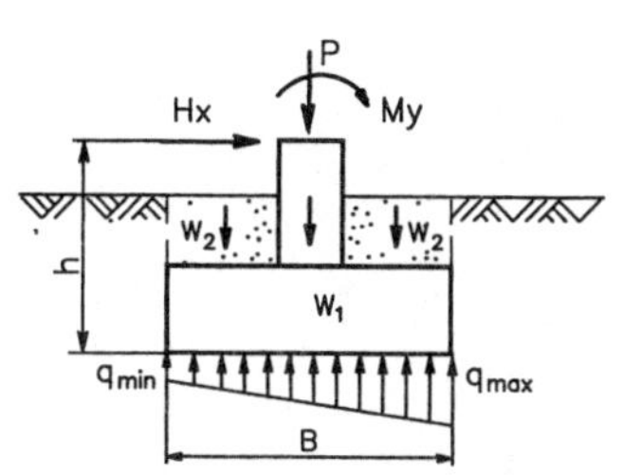

$$q_{max} = \frac{P_v}{A} + \frac{\sum M_y}{S_y}, \quad q_{min} = \frac{P_v}{A} - \frac{\sum M_y}{S_y}$$

Where $P_v = P + W_1 + 2W_2$

$$\sum M_y = H_x \cdot h + M_y$$

P = load on the footing from the column

W_1 = weight of concrete, including pedestal and base pad

W_2 = weight of soil

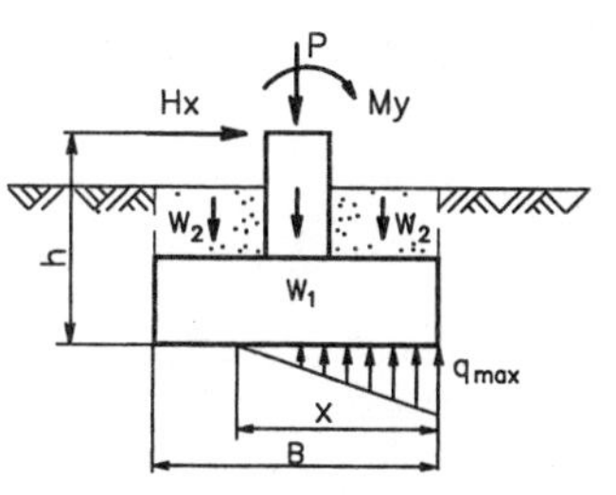

If $q_{min} < 0$, assume $q_{min} = 0$

(soil cannot furnish any tensile resistance)

$$x = \frac{3\left(P_v \cdot B - 2\sum M_y\right)}{2P_v}$$

$$q_{max} = \frac{2P_v}{x \cdot L}$$

N O T E S

Tables 9.1 and 9.2

Example. Direct foundation in Table 9.1

Given. Reinforced concrete footing, $B = 3.6\ \text{m},\quad L = 2.8\ \text{m},\quad h = 3\ \text{m}$

$A = B \cdot L = 3.6 \times 2.8 = 10.08\ \text{m}^2,\quad S_y = L \cdot B^2/6 = 6.048\ \text{m}^3$

Loads $P_v = P + W_1 + 2W_2 = 2250\ \text{kN},\quad M_y = 225\ \text{kN} \cdot \text{m},\quad H = 200\ \text{kN}$

Allowable soil contact pressure $\sigma = 360\ \text{kPa} = 360\ \text{kN/m}^2,\quad f = 0.4$

Required. Compute contact pressure, factors of safety against sliding and overturning

Solution. $q_{max} = \dfrac{P_v}{A} + \dfrac{\sum M_y}{S_y},\quad q_{min} = \dfrac{P_v}{A} - \dfrac{\sum M_y}{S_y}$

$$q_{max} = \frac{2250}{10.08} + \frac{200 \times 3 + 225}{6.048} = 223.2 + 136.4 = 359.6 < 360\ \text{kPa}$$

$$q_{min} = 223.2 - 136.4 = 86.8\ \text{kPa}$$

Factor of safety against sliding $\quad \text{F.S.} = \dfrac{P_v \cdot f}{\sum H} = \dfrac{2250 \times 0.4}{200} = 4.5$

Factor of safety against overturning $\quad \text{F.S.} = \dfrac{M_{r(k)}}{M_{0(k)}} = \dfrac{P_v \cdot B/2}{M + \sum H \cdot h} = \dfrac{2250 \times 3.6/2}{225 + 200 \times 3} = 4.9$

DIRECT FOUNDATION STABILITY

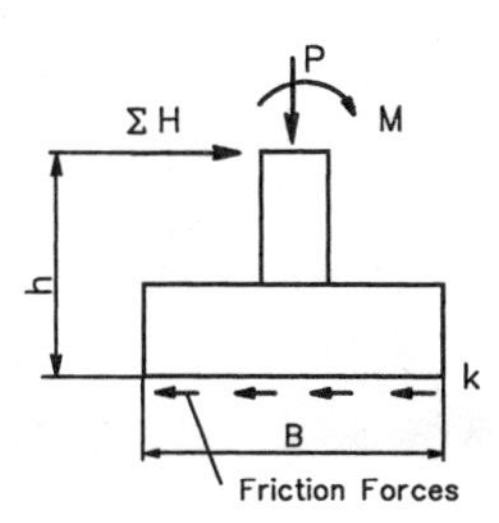

Factor of safety against sliding: $F.S. = \dfrac{P_v \cdot f}{\sum H}$

P_v = total vertical load, $\sum H$ = total horizontal forces

f = coefficient of friction between base and soil

$f \approx 0.4 - 0.5$

Factor of safety against overturning: $F.S. = \dfrac{M_{r(k)}}{M_{o(k)}}$

$M_{r(k)} = P_v \cdot B/2, \quad M_{o(k)} = M + \sum H \cdot h$

$M_{r(k)}$ = moment to resist turning

$M_{o(k)}$ = turning moment

PILE FOUNDATIONS

Distribution of loads in pile group

Example 9.2a

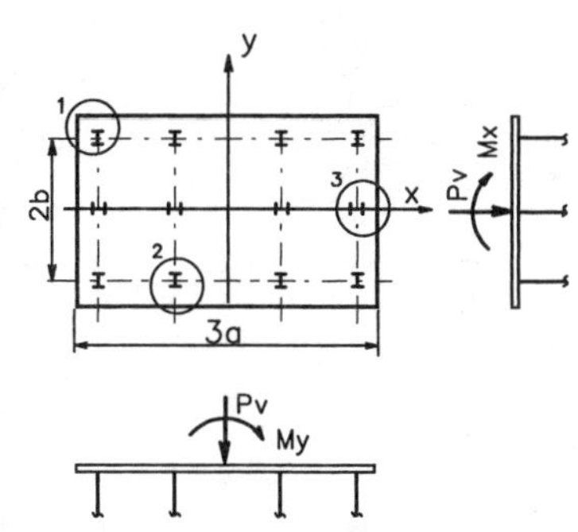

Foundation plan and sections

Axial load on any particular pile:

$$P_i = \frac{P_v}{n \cdot m} \pm \frac{M_y \cdot x}{\sum (x)^2} \pm \frac{M_x \cdot y}{\sum (y)^2}$$

P_v = total vertical load acting on pile group

n = number of piles in a row

m = number of rows of pile

M_x, M_y = moment with respect to x and y axes, respectively

x, y = distance from pile to y and x axes, respectively

Example 9.2a: $n = 4, \quad m = 3$

$\sum (x)^2 = 2 \cdot 3\left[(0.5a)^2 + (1.5a)^2\right] = 6 \cdot 6.25a = 13.5a$

$\sum (y)^2 = 2 \cdot 4 \cdot (b)^2 = 8b^2$

Pile 1: $x = -1.5a, \quad y = b, \quad P_1 = \dfrac{P_v}{4 \cdot 3} - \dfrac{M_y \cdot 1.5a}{13.5a^2} + \dfrac{M_x \cdot b}{8b^2} = \dfrac{P_v}{12} - \dfrac{M_y}{9a} + \dfrac{M_x}{8b}$

Pile 2: $x = -0.5a, \quad y = -b, \quad P_2 = \dfrac{P_v}{4 \cdot 3} - \dfrac{M_y \cdot 0.5a}{13.5a^2} - \dfrac{M_x \cdot b}{8b^2} = \dfrac{P_v}{12} - \dfrac{M_y}{27a} - \dfrac{M_x}{8b}$

Pile 3: $x = 0.5a, \quad y = 0, \quad P_3 = \dfrac{P_v}{4 \cdot 3} + \dfrac{M_y \cdot 1.5a}{13.5a^2} + \dfrac{M_x \cdot 0}{8b^2} = \dfrac{P_v}{12} + \dfrac{M_y}{9a}$

NOTES

Distribution of loads in pile group

Example 9.2b

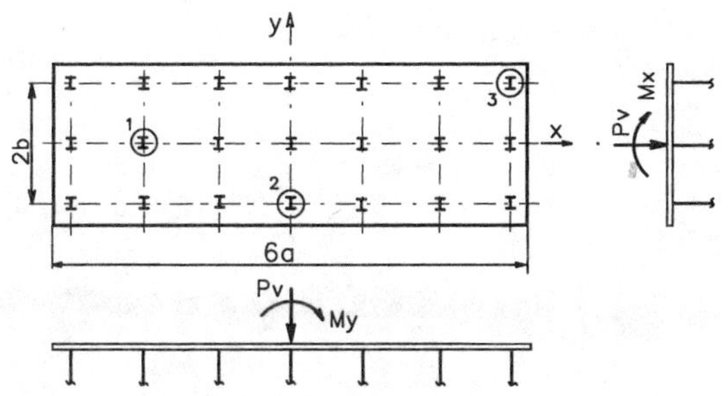

Foundation plan and sections

Axial load on any particular pile: $$P_i = \frac{P_v}{n \cdot m} \pm \frac{M_y \cdot x}{\sum(x)^2} \pm \frac{M_x \cdot y}{\sum(y)^2}$$

$n = 7, \quad m = 3, \quad \sum(x)^2 = 2 \cdot 3 \cdot \left[(a)^2 + (2a)^2 + (3a)^2\right] = 6 \cdot 14a^2 = 84a^2$

$\sum(y)^2 = 2 \cdot 7 \cdot (b)^2 = 14b^2$

Pile 1: $x = -2a, \quad y = 0, \quad P_1 = \frac{P_v}{7 \cdot 3} - \frac{M_y \cdot 2a}{84a^2} + \frac{M_x \cdot 0}{14b^2} = \frac{P_v}{21} - \frac{M_y}{42a}$

Pile 2: $x = 0, \quad y = -b, \quad P_2 = \frac{P_v}{7 \cdot 3} + \frac{M_y \cdot 0}{84a^2} - \frac{M_x \cdot b}{14b^2} = \frac{P_v}{21} - \frac{M_x}{14b}$

Pile 3: $x = 3a, \quad y = b, \quad P_3 = \frac{P_v}{7 \cdot 3} + \frac{M_y \cdot 3a}{84a^2} + \frac{M_x \cdot b}{14b^2} = \frac{P_v}{21} + \frac{M_y}{28a} + \frac{M_x}{14b}$

Maximum and minimum axial load on pile:

$$P_{\substack{max \\ min}} = \frac{P_v}{n \cdot m} \pm \frac{M_y}{S_x} \pm \frac{M_x}{S_y}, \quad S_x = \frac{n(n+1)a \cdot m}{6}, \quad S_y = \frac{m(m+1)b \cdot n}{6}$$

In example 9.2b : $S_x = \frac{7(7+1)a \cdot 3}{6} = 28a, \quad S_y = \frac{3(3+1)b \cdot 7}{6} = 14b$

PILE GROUP CAPACITY

$$N_g = E_g \cdot n \cdot m \cdot N_p$$

Converse-Labarre equation:

$$E_g = 1 - \left(\frac{\theta}{90}\right) \frac{(n-1)m + (m-1)n}{n \cdot m}$$

For cohesionless soil $E_g = 1.0$

Where N_g = capacity of the pile group

E_g = pile group efficiency

N_p = capacity of single pile

$\theta = \arctan d/s$ (degrees), d = diameter of piles,

s = min spacing of piles, center to center

NOTES

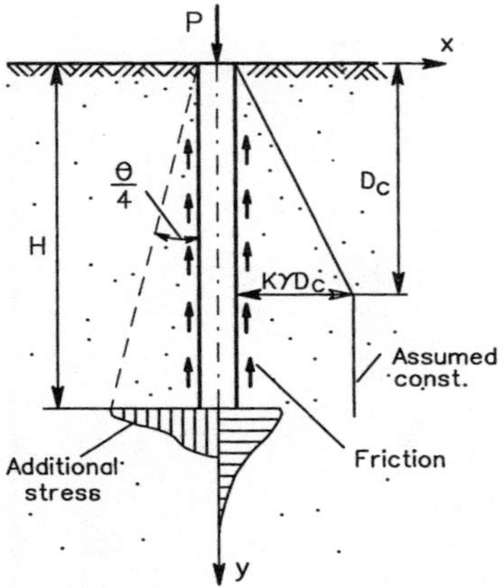

PILE–SOIL INTERACTION

θ = angle of internal friction

PILE CAPACITY

$$Q_u = Q_{fr} + Q_{tip}$$

Where: Q_u = ultimate (at failure) bearing capacity of a single pile

Q_{fr} = bearing capacity furnished by friction between the soil and the sides of pile

Q_{tip} = bearing capacity furnished by the soil just below the pile's tip

$$Q_{fr} = f_s \cdot C_p \left[0.5\gamma \cdot D_c^2 + \gamma \cdot D_c (H - D_c)\right] \cdot K, \qquad Q_{tip} = \gamma \cdot D_c \cdot N_q \cdot A_{tip}$$

Where: f_s = coefficient of friction between soil and pile.

Concrete: $f_s = 0.45$, wood: $f_s = 0.4$, steel: $f_s = 0.2 \div 0.4$

C_p = circumference of pile

γ = unit weight of soil

D_c = critical depth, ranging approximately from 10 pile diameters for loose sand to 20 pile diameters for dense compact sand

H = embedded length of pile

K = coefficient of lateral soil pressure

N_q = bearing capacity factor (see Table 8.7)

A_{tip} = area of the pile tip

NOTES

RIGID CONTINUOUS BEAM ELASTICALLY SUPPORTED

The following method can be applied on condition that : $L \le 0.8 \cdot h \cdot \sqrt[3]{E / E_s}$

Where E, L and h = modulus of elasticity, length and depth of the beam, respectively

E_s = modulus of deformation of soil

Uniformly distributed load (w)

1

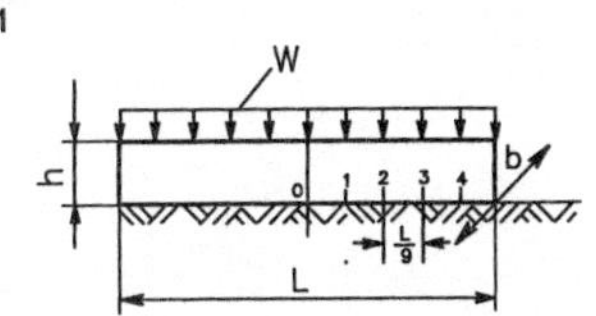

Soil reaction diagram

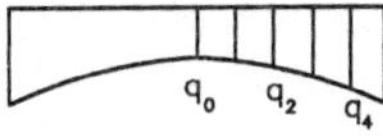

Moment diagram

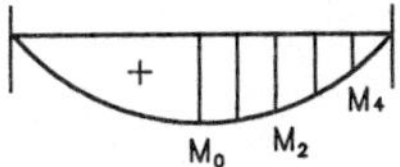

Shear diagram

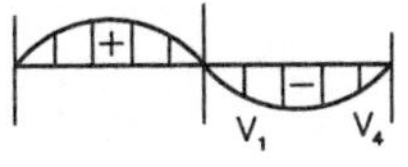

Soil reaction: $q_i = \alpha_{q(i)} \cdot w$

b/L	$\alpha_{q(0)}$	$\alpha_{q(1)}$	$\alpha_{q(2)}$	$\alpha_{q(3)}$	$\alpha_{q(4)}$
0.33	0.799	0.832	0.858	0.907	1.494
0.22	0.846	0.855	0.881	0.927	1.408
0.11	0.889	0.890	0.919	0.961	1.298
0.07	0.900	0.905	0.928	0.973	1.247

Bending moment: $M_i = \alpha_{m(i)} \cdot w \cdot b \cdot L^2$

b/L	$\alpha_{m(0)}$	$\alpha_{m(1)}$	$\alpha_{m(2)}$	$\alpha_{m(3)}$	$\alpha_{m(4)}$
0.33	0.018	0.014	0.010	0.006	0.001
0.22	0.012	0.011	0.009	0.005	0.001
0.11	0.009	0.008	0.006	0.004	0.000
0.07	0.008	0.007	0.006	0.003	0.000

Shear: $V_i = \alpha_{v(i)} \cdot w \cdot b \cdot L$

b/L	$\alpha_{v(0)}$	$\alpha_{v(1)}$	$\alpha_{v(2)}$	$\alpha_{v(3)}$	$\alpha_{v(5)}$
0.33	0.0	−0.019	−0.037	−0.050	−0.027
0.22	0.0	−0.016	−0.030	−0.041	−0.023
0.11	0.0	−0.014	−0.024	−0.031	−0.016
0.07	0.0	−0.012	−0.020	−0.026	−0.014

NOTES

RIGID CONTINUOUS BEAM ELASTICALLY SUPPORTED — 9.6

Concentrated loads

2

Moment diagram

Shear diagram

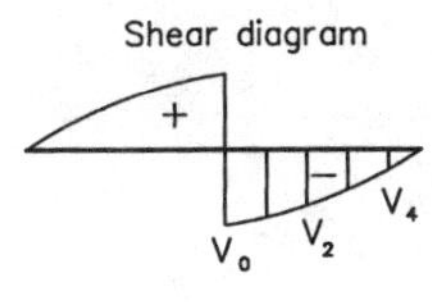

Bending moment: $M_i = \alpha_{m(i)} \cdot P \cdot L$

b/L	$\alpha_{m(0)}$	$\alpha_{m(1)}$	$\alpha_{m(2)}$	$\alpha_{m(3)}$	$\alpha_{m(4)}$
0.33	0.130	0.087	0.048	0.019	0.003
0.22	0.134	0.085	0.046	0.018	0.003
0.11	0.131	0.082	0.044	0.017	0.002
0.07	0.129	0.081	0.043	0.016	0.002

Shear: $V_i = \alpha_{v(i)} \cdot P$

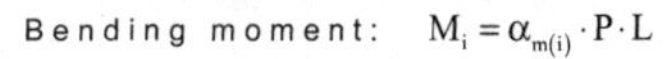

b/L	$\alpha_{v(0)}$	$\alpha_{v(1)}$	$\alpha_{v(2)}$	$\alpha_{v(3)}$	$\alpha_{v(5)}$
0.33	–0.500	–0.408	–0.314	–0.216	–0.083
0.22	–0.500	–0.404	–0.308	–0.208	–0.078
0.11	–0.500	–0.402	–0.302	–0.197	–0.072
0.07	–0.500	–0.400	–0.298	–0.192	–0.069

Moment diagram

Shear diagram

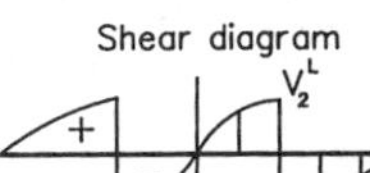

Bending moment: $M_i = \alpha_{m(i)} \cdot P \cdot L$

b/L	$\alpha_{m(0)}$	$\alpha_{m(1)}$	$\alpha_{m(2)}$	$\alpha_{m(3)}$	$\alpha_{m(4)}$
0.33	0.050	0.063	0.096	0.038	0.006
0.22	0.046	0.059	0.092	0.036	0.005
0.11	0.040	0.053	0.088	0.034	0.004
0.07	0.030	0.051	0.086	0.032	0.003

Shear: $V_i = \alpha_{v(i)} \cdot P$

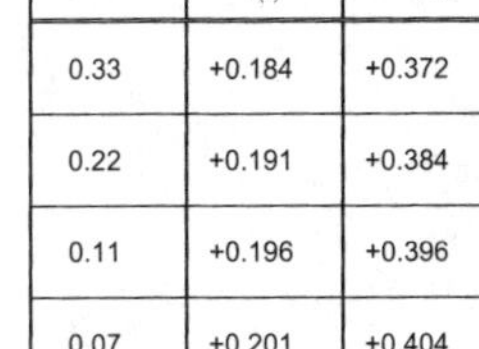

b/L	$\alpha_{v(1)}$	$\alpha^{L}_{v(2)}$	$\alpha^{R}_{v(2)}$	$\alpha_{v(3)}$	$\alpha_{v(4)}$
0.33	+0.184	+0.372	–0.628	–0.432	–0.166
0.22	+0.191	+0.384	–0.616	–0.416	–0.156
0.11	+0.196	+0.396	–0.604	–0.395	–0.144
0.07	+0.201	+0.404	–0.596	–0.385	–0.138

N O T E S

Table 9.7

Example. Rigid continuous footing 4 in Table 9.7

Given. Reinforced concrete footing, $L = 6$ m, $b = 2$ m, $h = 1$ m, $b/L = 0.33$

$E = 3370$ kip/in^2 $= 3370 \times 6.8948 = 23235$ MPa

$E_s = 40$ MPa, concentrated loads $P = 200$ kN

Required. Compute M_0, M_3, V_3^L, V_3^R

Solution. Checking condition: $L \leq 0.8 \cdot h \cdot \sqrt[3]{E/E_s}$, $6 < 0.8 \times 1 \times \sqrt[3]{23235/40} = 6.672$

$M_0 = \alpha_{m(0)} \times P \times L = -0.061 \times 200 \times 6 = -73.2 \text{ kN} \cdot \text{m}$

$M_3 = \alpha_{m(3)} \times P \times L = 0.038 \times 200 \times 6 = 45.6 \text{ kN} \cdot \text{m}$

$V_3^L = \alpha_{v(3)} \times P = 0.568 \times 200 = 113.6 \text{ kN}$

$V_3^R = -0.432 \times 200 = -86.4 \text{ kN}$

FOUNDATIONS

RIGID CONTINUOUS BEAM ELASTICALLY SUPPORTED

9.7

Concentrated loads

4

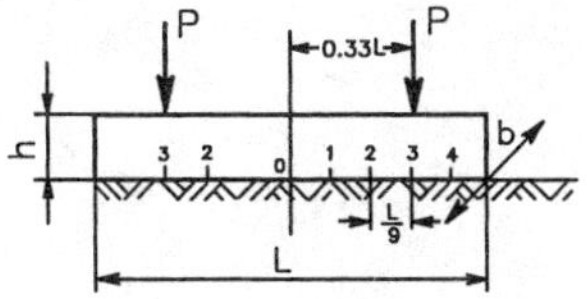

Moment diagram

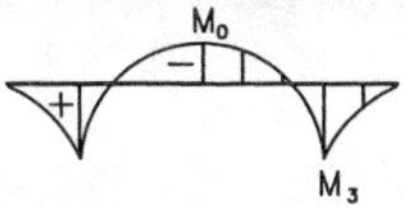

Shear diagram

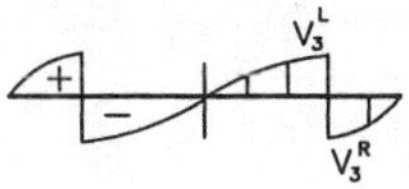

Bending moment: $M_i = \alpha_{m(i)} \cdot P \cdot L$

b/L	$\alpha_{m(0)}$	$\alpha_{m(1)}$	$\alpha_{m(2)}$	$\alpha_{m(3)}$	$\alpha_{m(4)}$
0.33	–0.061	–0.048	–0.015	+0.038	+0.006
0.22	–0.065	–0.052	–0.019	+0.036	+0.005
0.11	–0.071	–0.058	–0.023	+0.034	+0.004
0.07	–0.075	–0.060	–0.025	+0.032	+0.004

Shear: $V_i = \alpha_{v(i)} \cdot P$

b/L	$\alpha_{v(1)}$	$\alpha_{v(2)}$	$\alpha^{L}_{v(3)}$	$\alpha^{R}_{v(3)}$	$\alpha_{v(4)}$
0.33	+0.184	+0.372	+0.568	–0.432	–0.166
0.22	+0.191	+0.384	+0.584	–0.416	–0.156
0.11	+0.196	+0.396	+0.605	–0.395	–0.144
0.07	+0.211	+0.404	+0.615	–0.385	–0.138

5

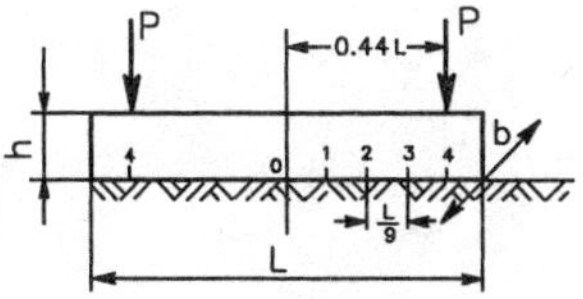

Moment diagram

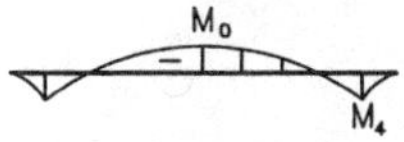

Shear diagram

Bending moment: $M_i = \alpha_{m(i)} \cdot P \cdot L$

b/L	$\alpha_{m(0)}$	$\alpha_{m(1)}$	$\alpha_{m(2)}$	$\alpha_{m(3)}$	$\alpha_{m(4)}$
0.33	–0.172	–0.159	–0.126	–0.073	+0.006
0.22	–0.176	–0.163	–0.130	–0.075	+0.005
0.11	–0.182	–0.169	–0.134	–0.077	+0.004
0.07	–0.186	–0.171	–0.136	–0.079	+0.004

Shear: $V_i = \alpha_{v(i)} \cdot P$

b/L	$\alpha_{v(1)}$	$\alpha_{v(2)}$	$\alpha_{v(3)}$	$\alpha^{L}_{v(4)}$	$\alpha^{R}_{v(4)}$
0.33	+0.184	+0.372	+0.568	+0.834	–0.166
0.22	+0.191	+0.384	+0.584	+0.844	–0.156
0.11	+0.196	+0.396	+0.605	+0.856	–0.144
0.07	+0.201	+0.404	+0.615	+0.862	–0.138

NOTES

10, 11. RETAINING STRUCTURES

N O T E S

For determining the lateral earth pressure on walls of structures, the methods that have proved most popular in engineering practice are those based on analysis of the sliding prism's standing balance. The magnitude of the lateral earth pressure is dependent on the direction of the wall movement. This correlation is represented graphically in Table 10.1. The three known coordinates on the graph are P_a, P_0 and P_p. As the graph demonstrates, the active pressure is the smallest, and the passive pressure the largest, among the forces and reactions acting between the soil and the wall.
Construction experience shows that even a minor movement of the retaining walls away from the soil in many cases leads to the formation of a sliding prism and produces active lateral pressure.

RETAINING STRUCTURES

LATERAL EARTH PRESSURE ON RETAINING WALLS

10.1

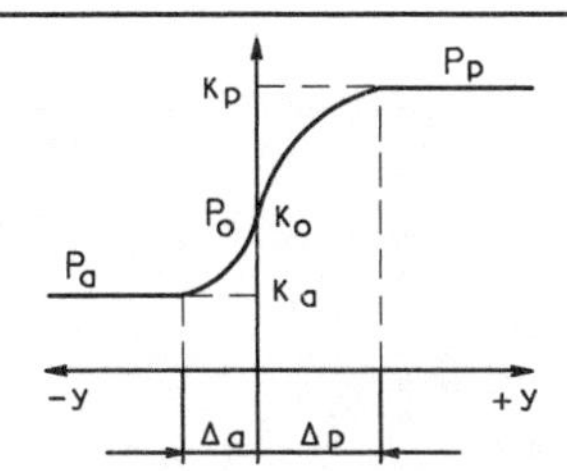

Correlation between lateral earth pressure and wall movement

P_0 = lateral earth pressure at rest

P_a = active lateral earth pressure

P_p = passive lateral earth pressure

K_0, K_a, K_p = coefficients

Coefficients of lateral earth pressure:

K_0 = coefficient of earth pressure at rest: $K_0 = \dfrac{\sigma_h}{\sigma_v} = \dfrac{\mu}{1-\mu}$

Where σ_h and σ_v = lateral and vertical stresses, respectively

μ = Poisson's ratio

Type of soil	μ
Sand	0.29
Sandy loam	0.31
Sandy clay	0.37
Clay	0.41

Alternative formulas: $K_0 = 1 - \sin\phi$ - for sands

$K_0 = 0.19 + 0.233\log(PI)$ - for clays

Where PI = soil's plasticity index

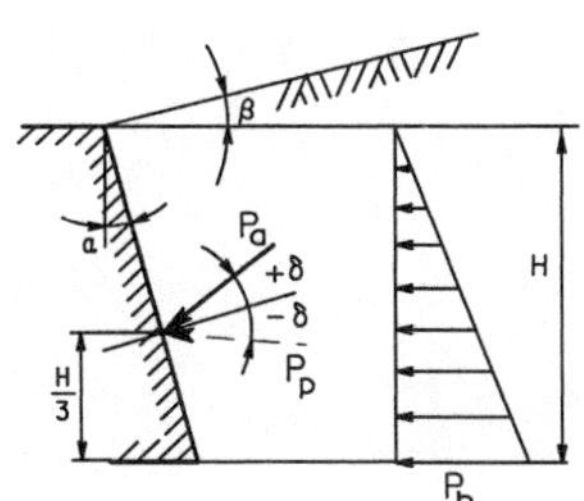

Coulomb earth pressure

$P_a = 0.5K_a\gamma H^2, \quad P_p = 0.5K_p\gamma H^2$

Where γ = unit weight of the backfill soil

K_a = coefficient of active earth pressure

K_p = coefficient of passive earth pressure

Coulomb theory

$$K_a = \frac{\cos^2(\phi-\alpha)}{\left[1+\sqrt{\dfrac{\sin(\phi+\delta)\sin(\phi-\beta)}{\cos(\alpha+\delta)\cos(\beta-\alpha)}}\right]^2 \cos^2\alpha\cdot\cos(\alpha+\delta)}$$

$$K_p = \frac{\cos^2(\phi-\alpha)}{\left[1-\sqrt{\dfrac{\sin(\phi+\delta)\sin(\phi+\beta)}{\cos(\alpha-\delta)\cos(\beta-\alpha)}}\right]^2 \cos^2\alpha\cdot\cos(\alpha-\delta)}$$

Where : ϕ = angle of internal friction of the backfill soil

δ = angle of friction between wall and soil $(\delta \approx 2/3\phi)$

β = angle between backfill surface line and a horizontal line

α = angle between back side of wall and a vertical line

EARTHQUAKE

$$K_{aE} = \frac{\cos^2(\phi-\theta-\alpha)}{\left[1+\sqrt{\dfrac{\sin(\phi+\delta)\sin(\phi-\theta-\beta)}{\cos(\alpha+\delta+\theta)\cos(\beta-\alpha)}}\right]^2 \cos\theta\cdot\cos^2\alpha\cdot\cos(\alpha+\theta+\delta)}$$

$\theta = \arctan\left[k_h/(1-k_v)\right]$

k_h = seismic coefficient, $k_h = A_E/2$

A_E = acceleration coefficient

k_v = vertical acceleration coefficient

NOTES

Table 10.2

Example. Retaining wall 1 in Table 10.2, $H = 10\ \text{m}$

Given. Cohesive soil, angle of friction $\phi = 26^0$

Cohesion $c = 150\ \text{lb/ft}^2 = 150 \times 47.88 = 7182\ \text{Pa} = 7.2\ \text{kN/m}^2$

Unit weight of backfill soil $\gamma = 115\ \text{lb/ft}^3 = 115 \times 0.1571 = 18.1\ \text{kN/m}^3$

Required. Compute active and passive earth pressure per unit length of wall: P_a, h, P_p, d_p

Solution. Active earth pressure:

$$K_a = \tan^2\left(45^0 - \frac{\phi}{2}\right) = \tan^2\left(45^0 - \frac{26^0}{2}\right) = 0.39$$

$$p_h = K_a \gamma H - 2c\sqrt{K_a} = 0.39 \times 18.1 \times 10 - 2 \times 7.2\sqrt{0.39} = 61.61\ \text{kN/m}$$

$$h = \frac{p_h H}{p_h + 2c\tan\left(45^0 - \frac{\phi}{2}\right)} = \frac{61.61 \times 10}{61.61 + 2 \times 7.2 \times 0.624} = 8.73\ \text{m}$$

$$P_a = 0.5 p_h h = 0.5 \times 61.61 \times 8.73 = 269\ \text{kN}$$

Passive earth pressure:

$$K_p = \tan^2\left(45^0 + \frac{\phi}{2}\right) = \tan^2\left(45^0 + \frac{26^0}{2}\right) = 2.56$$

$$p_h = K_p \gamma H + 2c\sqrt{K_p} = 2.56 \times 18.1 \times 10 + 2 \times 7.2\sqrt{2.56} = 486.4\ \text{kN/m}$$

$$P_p = 0.5\left[2c\tan\left(45^0 + \frac{\phi}{2}\right) + p_h\right]H = 0.5[23.04 + 486.4] \times 10 = 2547.2\ \text{kN}$$

$$d_p = \frac{p_h + 4c\tan\left(45^0 + \frac{\phi}{2}\right)}{3\left[p_h + 2c\tan\left(45^0 + \frac{\phi}{2}\right)\right]}H = \frac{486.4 + 4 \times 7.2 \times 1.6}{3[486.4 + 2 \times 7.2 \times 1.6]} \times 10 = 3.48\ \text{m}$$

LATERAL EARTH PRESSURE ON RETAINING WALLS

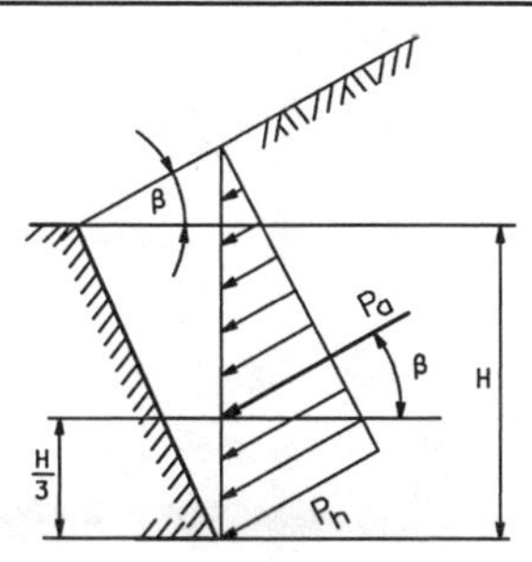

Rankine earth pressure

$P_a = 0.5K_a\gamma H^2 \quad P_p = 0.5K_p\gamma H^2$

Rankine theory $(\alpha = 0,\ \delta = 0)$

The wall is assumed to be vertical and smooth

$$K_a = \cos\beta \frac{\cos\beta - \sqrt{\cos^2\beta - \cos^2\phi}}{\cos\beta + \sqrt{\cos^2\beta - \cos^2\phi}}$$

$$K_p = \cos\beta \frac{\cos\beta + \sqrt{\cos^2\beta - \cos^2\phi}}{\cos\beta - \sqrt{\cos^2\beta - \cos^2\phi}}$$

If $\alpha = 0,\ \delta = 0$ and $\beta = 0$:

$$K_a = \frac{1-\sin\phi}{1+\sin\phi} = \tan^2\left(45^0 - \frac{\phi}{2}\right)$$

$$K_p = \frac{1+\sin\phi}{1-\sin\phi} = \tan^2\left(45^0 + \frac{\phi}{2}\right) = \frac{1}{K_a}$$

Examples

1. Assumed: $\alpha = 0,\ \delta = 0,\ \beta = 0$

A /

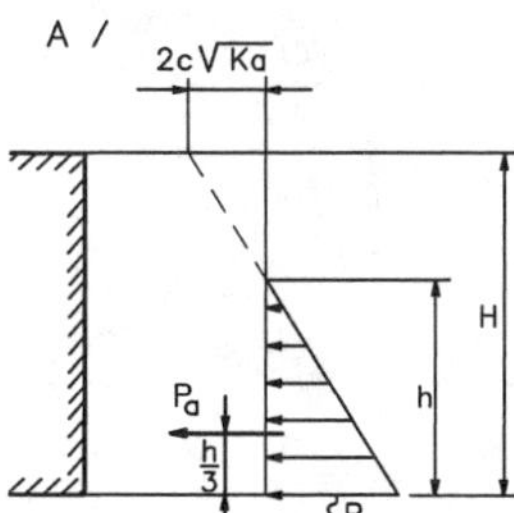

B /

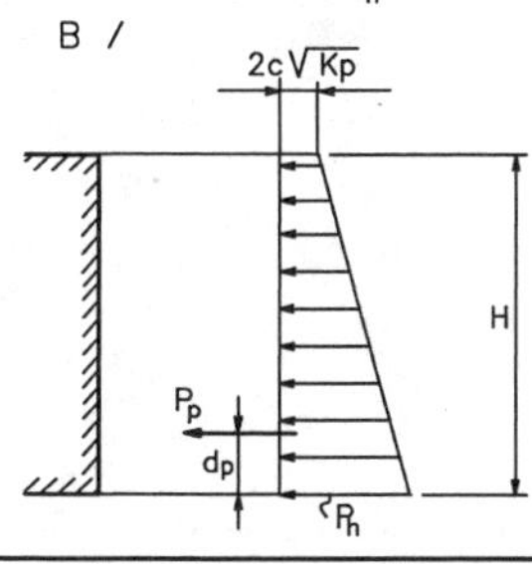

Cohesive soil

A/ **Active earth pressure**

$$p_h = K_a\gamma H - 2c\sqrt{K_a}$$

Where c = unit cohesive strength of soil

$$K_a = \tan^2\left(45^0 - \frac{\phi}{2}\right), \qquad h = \frac{p_h \cdot H}{p_h + 2c\tan\left(45^0 - \frac{\phi}{2}\right)}$$

Resultant force per unit length of wall $\quad P_a = 0.5 p_h h$

B/ **Passive earth pressure**

$$p_h = K_p\gamma H + 2c\sqrt{K_p}, \qquad K_p = \tan^2\left(45^0 + \frac{\phi}{2}\right)$$

$$P_p = 0.5\left[2c\tan\left(45^0 + \frac{\phi}{2}\right) + p_h\right]\cdot H$$

$$d_p = \frac{p_h + 4c\cdot\tan\left(45^0 + \frac{\phi}{2}\right)}{3\left[p_h + 2c\cdot\tan\left(45^0 + \frac{\phi}{2}\right)\right]}\cdot H$$

N O T E S

Table 10.3

Example. Retaining wall 3 in Table 10.3, $H = 6$ m

Given. Backfill soil: Angle of friction $\phi = 30^0$, cohesion $c = 0$

Unit weight of backfill soil $\gamma = 18$ kN/m^3

Ground water: $h_w = 4$ m, $\gamma_w = 9.81$ kN/m^3

Required. Compute active pressure per unit length of wall: P_a, d_a

Solution.

$$K_a = \tan^2\left(45^0 - \frac{\phi}{2}\right) = \tan^2\left(45^0 - \frac{30^0}{2}\right) = 0.333$$

$$P_1 = 0.5K_a\gamma(H-h_w)^2 = 0.5\times0.333\times18(6-4)^2 = 12.0 \text{ kN}$$

$$d_1 = \frac{H-h_w}{3} + h_w = \frac{6-4}{3} + 4 = 4.67 \text{ m}$$

$$P_2 = K_a\gamma(H-h_w)h_w = 0.333\times18(6-4)\times4 = 48.0 \text{ kN}$$

$$d_2 = 0.5h_w = 0.5\times4 = 2 \text{ m}$$

$$P_3 = 0.5K_a(\gamma-\gamma_w)h_w^2 = 0.5\times0.333\times(18-9.81)\times4^2 = 21.8 \text{ kN}$$

$$d_3 = \frac{h_w}{3} = \frac{4}{3} = 1.33 \text{ m}$$

$$P_4 = 0.5\gamma_w h_w^2 = 0.5\times9.81\times4^2 = 78.5 \text{ kN}$$

$$d_4 = \frac{h_w}{3} = \frac{4}{3} = 1.33 \text{ m}$$

$$P_a = P_1 + P_2 + P_3 + P_4 = 12.0 + 48.0 + 21.8 + 78.5 = 160.3 \text{ kN}$$

$$d_a = \frac{P_1d_1 + P_2d_2 + P_3d_3 + P_4d_4}{P_a} = \frac{12.0\times4.67 + 48.0\times2 + 21.8\times1.33 + 78.5\times1.33}{160.3} = 1.78 \text{ m}$$

2

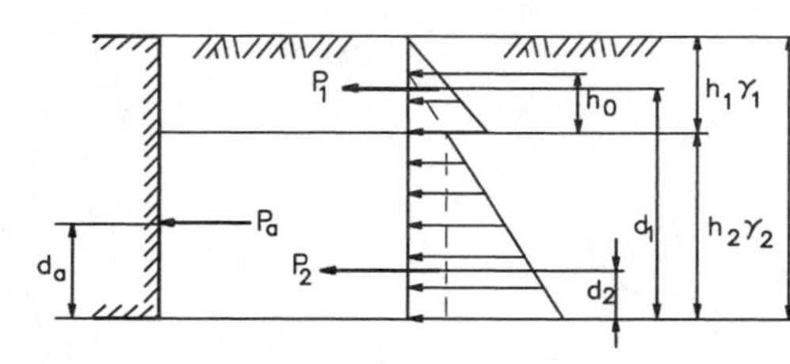

Active earth pressure

$$P_1 = 0.5K_a\gamma_1 h_1^2, \quad d_1 = h_2 + \frac{h_1}{3}, \quad h_0 = \frac{\gamma_1 h_1}{\gamma_2}$$

$$P_2 = 0.5K_a\gamma_2\left(2h_0 + h_2\right)h_2, \quad d_2 = \frac{h_2 + 3h_0}{h_2 + 2h_0}\cdot\frac{h_2}{3}$$

Total active earth pressure $P_a = P_1 + P_2$

$$d_a = \frac{P_1 d_1 + P_2 d_2}{P_a}$$

3

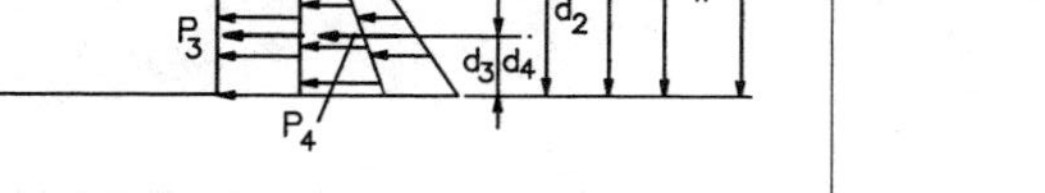

Active earth pressure

$$P_1 = 0.5K_a\gamma\left(H - h_w\right)^2, \quad d_1 = \frac{H - h_w}{3} + h_w$$

$$P_2 = K_a\gamma\left(H - h_w\right)h_w, \quad d_2 = 0.5h_w$$

$$P_3 = 0.5K_a\left(\gamma - \gamma_w\right)h_w^2, \quad d_3 = \frac{h_w}{3}$$

γ_w = unit weight of water ($\gamma_w = 9.81\ kN/m^3$)

$$P_4 = 0.5\gamma_w h_w^2, \quad d_4 = \frac{h_w}{3}$$

Total active earth pressure $P_a = P_1 + P_2 + P_3 + P_4$

$$d_a = \frac{P_1 d_1 + P_2 d_2 + P_3 d_3 + P_4 d_4}{P_a}$$

NOTES

4

Active earth pressure

$$p_h = K_a \gamma H, \quad P_1 = 0.5 K_a \gamma H^2, \quad d_1 = \frac{H}{3}$$

$$q_h = K_a w, \quad P_2 = 0.5 K_a w H, \quad d_2 = \frac{H}{2}$$

w = uniformly distributed load

Total active earth pressure $P_a = P_1 + P_2$

$$d_a = \frac{P_1 d_1 + P_2 d_2}{P_a} = \frac{H + 3w/\gamma}{H + 2w/\gamma} \cdot \frac{H}{3}$$

5

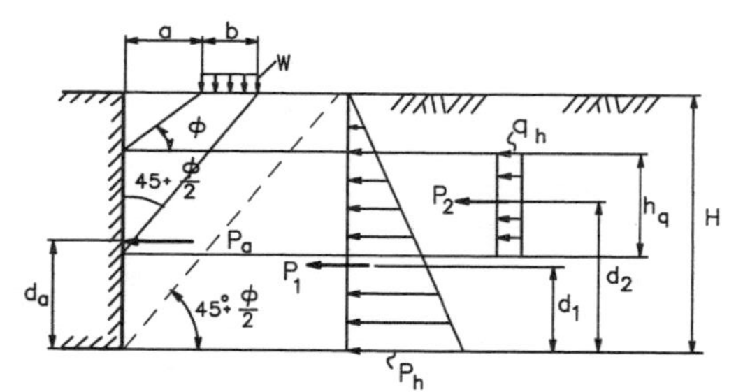

Active earth pressure

$$p_h = K_a \gamma H, \quad P_1 = 0.5 K_a \gamma H^2, \quad d_1 = \frac{H}{3}$$

$$q_h = K_a w, \quad P_2 = K_a w h_q,$$

$$h_q = (a+b)\tan\left(45^0 + \frac{\phi}{2}\right) - a\tan\phi$$

$$d_2 = H - \frac{1}{2}\left[(a+b)\tan\left(45^0 + \frac{\phi}{2}\right) + a\tan\phi\right]$$

Total active earth pressure $P_a = P_1 + P_2$

$$d_a = \frac{P_1 d_1 + P_2 d_2}{P_a}$$

NOTES

6

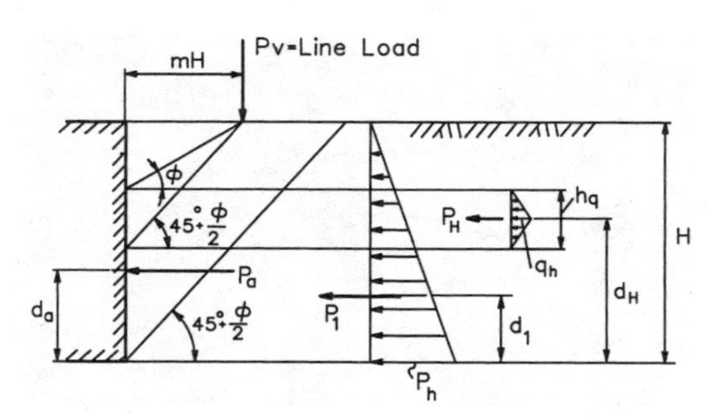

Active earth pressure

$$p_h = K_a \gamma H, \quad P_1 = 0.5 K_a \gamma H^2, \quad d_1 = \frac{H}{3}$$

$$q_h = 2K_a \frac{P_v}{mH}\cos\phi, \quad P_H = 0.5 q_h h_q$$

$$h_q = mH\left[\tan\left(45^0 + \frac{\phi}{2}\right) - \tan\phi\right]$$

$$d_H = H - \left(0.5 h_q + mH\tan\phi\right)$$

Total active earth pressure $P_a = P_1 + P_H$

$$d_a = \frac{P_1 d_1 + P_H d_H}{P_a}$$

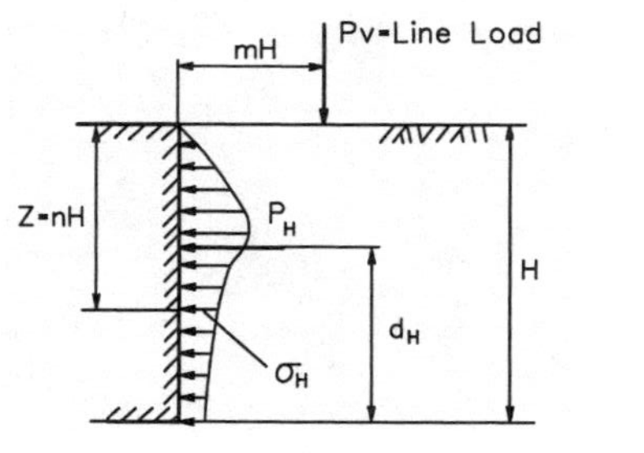

Alternative formulas

For $m \le 0.4$ $\quad \sigma_H = 0.20\frac{P_v}{H}\cdot\frac{n}{\left(0.16+n^2\right)^2}$

$P_H = 0.55 P_v$, $\quad d_H = 0.60H$

For $m > 0.4$ $\quad \sigma_H = 1.28\frac{P_v}{H}\cdot\frac{m^2 n}{\left(m^2+n^2\right)^2}$

$P_H = \frac{0.64 P_v}{\left(m^2+1\right)}$, $\quad d_H = 0.56H \quad (\text{For } m = 0.5)$

$d_H = 0.48H \quad (\text{For } m = 0.7)$

NOTES

RETAINING STRUCTURES

LATERAL EARTH PRESSURE ON BRACED SHEETINGS

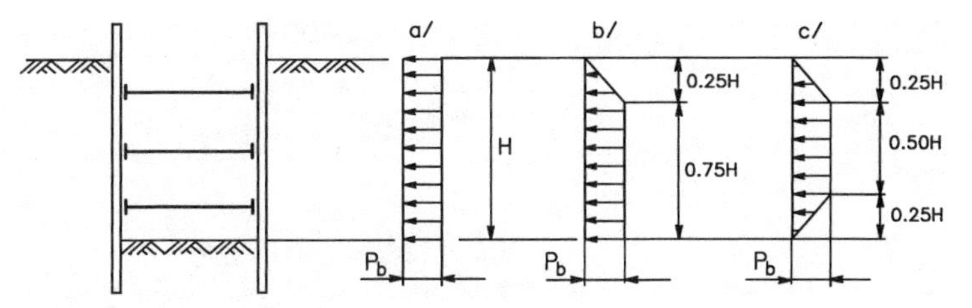

Empirical diagrams of lateral earth pressure on braced sheetings

a / Sand: $p_b = 0.65\gamma H \tan^2\left(45^0 - \frac{\phi}{2}\right)$

b / Soft to medium clay: $p_b = \gamma H - 2q_u$, q_u = unconfined compressive strength, $q_u = 2c$

c / Stiff-fissured clay: $p_b = 0.2\gamma H$ to $0.4\gamma H$

LATERAL EARTH PRESSURE ON BASEMENT WALLS

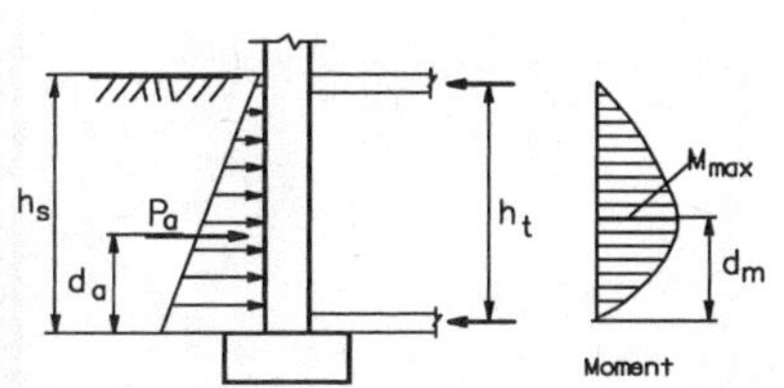

Active earth pressure:

$$P_a = 0.5K\gamma h_s^2 \ , \quad d_a = \frac{h_s}{3}$$

Maximum bending moment:

$$M_{max} = 0.128 P_a h_s \ , \quad d_m = 0.42 h_s$$

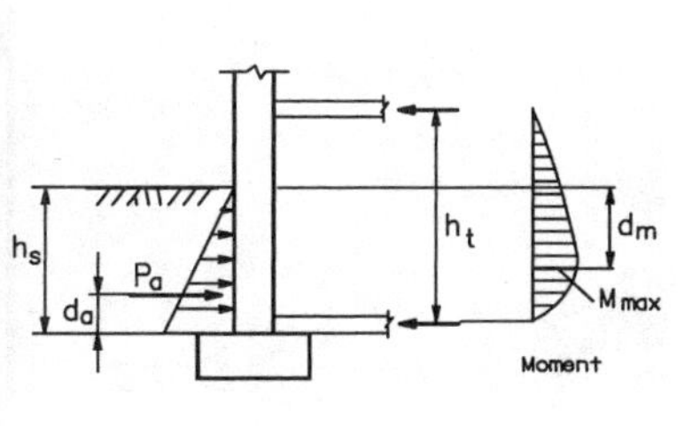

Active earth pressure:

$$P_a = 0.5K_a\gamma h_s^2 \ , \quad d_a = \frac{h_s}{3}$$

Maximum bending moment:

$$M_{max} = \frac{P_a h_s}{3h_t}\left(h + \frac{2h_s}{3}\sqrt{\frac{h_s}{3h_t}}\right) , \quad d_m = h_s\sqrt{\frac{h_s}{3h_t}}$$

NOTES

CANTILEVER RETAINING WALLS

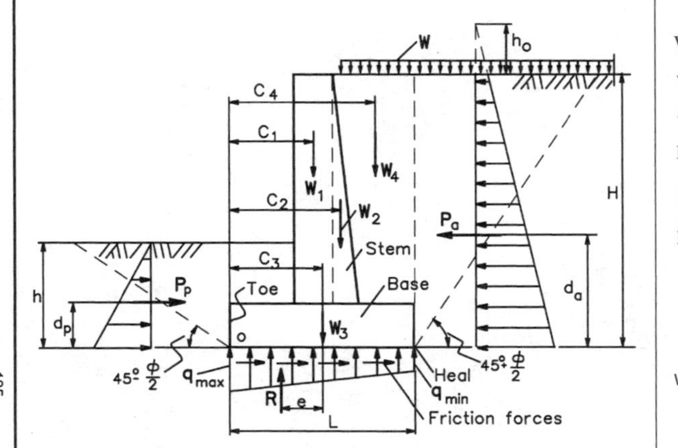

Stability analysis

W_i = weight (concentrated load for width $B = 1$)

w = surcharge (uniformly distributed load), $h_0 = w / \gamma$

Active earth pressure :

$$P_a = 0.5\gamma H \tan^2\left(45^0 - \frac{\phi}{2}\right)(H + 2h_0)\ ,\ d_a = \frac{H}{3} \cdot \frac{H + 3h_0}{H + 2h_0}$$

Passive earth pressure :

$$P_p = 0.5\gamma h \tan^2\left(45^0 + \frac{\phi}{2}\right)\ ,\quad d_p = \frac{h}{3}$$

The factor of safety against sliding

$$F.S. = \frac{\text{resisting force F}}{\text{actual horizontal force} \sum P_H}$$

Where $F = f\sum W_i$, $\sum P_H = P_a - P_p$

f = coefficient of friction ($f = 0.4$ to 0.5)

$F.S. = 1.5$ to 2.0

The factor of safety against overturning

$$F.S. = \frac{\text{Stabilizing moment about toe } \left(\sum M_r\right)}{\text{Overturning moment about toe } \left(\sum M_o\right)}$$

Where $\sum M_r = \sum W_i c_i + P_p d_p$, $\sum M_o = P_a d_a$

$F.S. = 1.5$ to 2.0

The factor of safety of bearing capacity failure

$$F.S. = \frac{\text{Soil's ultimate bearing capacity}}{\text{Maximum contact (base) pressure}}\ ,\qquad F.S. = 3.0$$

Eccentricity of resultant force R : $e = \frac{L}{2} - \frac{\sum M_o}{\sum W_i} \le \frac{L}{6}$, $R = \sum W_i$

Maximum contact (base) pressure : $q_{max} = \frac{\sum W_i}{L \cdot B} + \frac{6\sum W_i \cdot e}{L^2 \cdot B}$, (B=1)

N O T E S

Table 11.2

Example. Cantilever sheet piling 2 in Table 11.2, $H = 10\ \text{m}$

Given. Soil properties: $\phi_1 = 32^0, \quad c_1 = 0, \quad \gamma_1 = 18\ \text{kN/m}^3$

$\phi_2 = 34^0, \quad c_2 = 0, \quad \gamma_2 = 16\ \text{kN/m}^3, \qquad \beta = 0, \quad \alpha = 0, \quad \delta = 0$

Required. Compute depth D and maximum bending moment M_{max} per unit length of sheet piling

Solution. $K_{a_1} = \tan^2\left(45^0 - \frac{\phi_1}{2}\right) = \tan^2\left(45^0 - \frac{32^0}{2}\right) = 0.307$

$$K_{a_2} = \tan^2\left(45^0 - \frac{\phi_2}{2}\right) = \tan^2\left(45^0 - \frac{34^0}{2}\right) = 0.283$$

$$K_{p_2} = \tan^2\left(45^0 + \frac{\phi_2}{2}\right) = \tan^2\left(45^0 + \frac{34^0}{2}\right) = 3.537, \qquad K_{p_2} - K_{a_2} = 3.254$$

$$P_1 = 0.5K_{a_1}\gamma_1 H^2 = 0.5 \times 0.307 \times 18 \times 10^2 = 276.3\ \text{kN},$$

$$z_1 = \frac{K_{a_2}\gamma_1 H}{\left(K_{p_2} - K_{a_2}\right)\gamma_2} = \frac{0.283 \times 18 \times 10}{3.254 \times 16} = 0.98\ \text{m}$$

$$P_2 = 0.5K_{a_2}\gamma_1 H z_1 = 0.5 \times 0.283 \times 18 \times 10 \times 0.98 = 24.96\ \text{kN},$$

$$z_2 = \sqrt{\frac{P_1 + P_2}{0.5\left(K_{p_2} - K_{a_2}\right)\gamma_2}} = \sqrt{\frac{276.3 + 24.96}{0.5 \times 3.254 \times 16}} = 3.4\ \text{m}$$

$$P_3 = 0.5\left(K_{p_2} - K_{a_2}\right)\gamma_2\left(D_0 - z_1\right)^2 = 0.5 \times 3.254 \times 16\left(D_0 - z_1\right)^2 = 26.03\left(D_0 - z_1\right)^2$$

$\sum M_d = 0$ (condition of equilibrium)

$$P_1\left(\frac{H}{3} + D_0\right) + P_2\left(D_0 - \frac{z_1}{3}\right) - P_3\frac{1}{3}\left(D_0 - z_1\right) = 0$$

$$276.3\left(\frac{10}{3} + D_0\right) + 24.96D_0 - 26.03\frac{1}{3}\left(D_0 - z_1\right)^3 = 0$$

$$8.68\left(D_0 - z_1\right)^3 = 921.0 + 301.26D_0$$

Using method of trial and error:

assume $D_0 = 8.3\ \text{m}$, $(8.3 - 0.98)^3 = 106.10 + 34.71 \times 8.3, \quad 394.19 \approx 393.18$

$$D = 1.2D_0 = 1.2 \times 8.3 = 9.96\ \text{m}$$

$$M_{max} = P_1\left(\frac{H}{3} + z_1 + z_2\right) + P_2\left(\frac{2}{3}z_1 + z_2\right) - 0.5\left(K_{p_2} - K_{a_2}\right)\gamma_2 z_2^2\left(\frac{z_2}{3}\right)$$

$$= 276.3\left(\frac{10}{3} + 0.98 + 3.4\right) + 24.96\left(\frac{2}{3} \times 0.98 + 3.4\right) - 0.5 \times 3.254 \times 16 \times 3.4^2\left(\frac{3.4}{3}\right) = 1891.4\ \text{kN} \cdot \text{m/m}$$

1

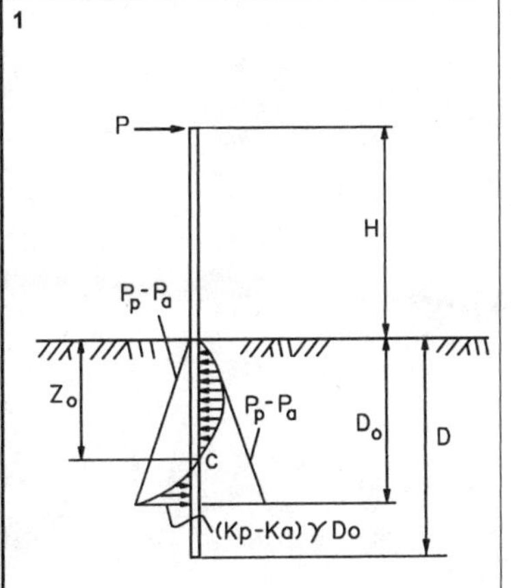

Equation to determine the embedment (D_0):

$$P = \frac{(K_p - K_a)\gamma D_0^3}{6(4H + 3D_0)}$$

Maximum bending moment :

$$M_{max} = P\left(H + \frac{2}{3}\sqrt{\frac{P}{(K_p - K_a)\gamma}}\right)$$

$$z_c = D_0 \frac{4H + 3D_0}{6H + 4D_0}$$

For single pile

$$P = \frac{(K_p - K_a)\gamma d D_0^3}{3(4H + 3D_0)},\quad M_{max} = P\left(H + \frac{2}{3}\sqrt{\frac{P}{(K_p - K_a)\gamma d}}\right),$$

where d = pile diameter

$D = (1.2 \text{ to } 1.4)D_0$ for factor of safety at 1.5 to 2.0

2

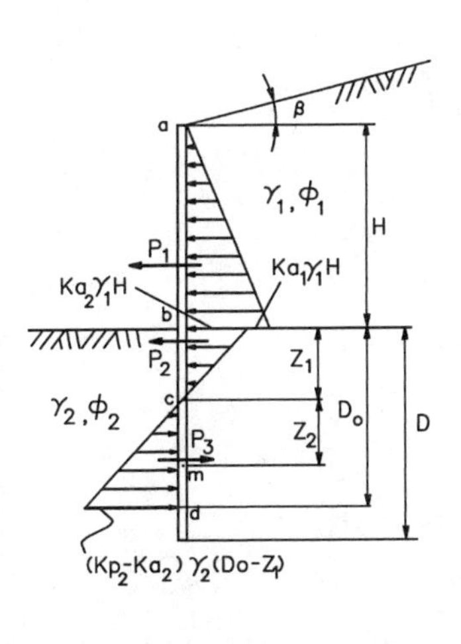

Earth pressure:

$$P_1 = 0.5K_a\gamma H^2, \qquad z_1 = \frac{K_a\gamma_1 H}{(K_{p_2} - K_{a_2})\gamma_2}$$

$$P_2 = 0.5K_{a_2}\gamma_1 H \cdot z_1, \qquad z_2 = \sqrt{\frac{P_1 + P_2}{0.5(K_{p_2} - K_{a_2})\gamma_2}}$$

$$P_3 = 0.5(K_{p_2} - K_{a_2})\gamma_2(D_0 - z_1)^2$$

Equation to determine D_0: $\sum M_d = 0$

$$P_1\left(\frac{H}{3} + D_0\right) + P_2\left(D_0 - \frac{z_1}{3}\right) - P_3\frac{1}{3}(D_0 - z_1) = 0$$

$D = (1.2 \text{ to } 1.4)D_0$ for factor of safety at 1.5 to 2.0

m = point of zero shear and maximum bending moment

Maximum bending moment

$$M_{max} = P_1\left(\frac{H}{3} + z_1 + z_2\right) + P_2\left(\frac{2}{3}z_1 + z_2\right)$$
$$-0.5(K_p - K_a)\gamma z_2^2 \cdot \left(\frac{z_2}{3}\right)$$

NOTES

Table 11.3

Example. Anchored sheet pile wall in Table 11.3, $H = 15\ \text{m}$

Given. Soil properties: $\phi_1 = 30^0,\ c_1 = 0,\ \gamma_1 = 20\ \text{kN/m}^3,\ \phi_2 = 32^0,\ c_2 = 0,\ \gamma_2 = 18\ \text{kN/m}^3$

$\beta = 0,\ \alpha = 0,\ \delta = 0,\ d = 1.2\ \text{m}$

Required. Compute depth D and maximum bending moment M_{max} per unit length of wall

Solution.

$$K_{a_1} = \tan^2\left(45^0 - \frac{\phi_1}{2}\right) = \tan^2\left(45^0 - \frac{30^0}{2}\right) = 0.333,\quad K_{a_2} = \tan^2\left(45^0 - \frac{\phi_2}{2}\right) = \tan^2\left(45^0 - \frac{32^0}{2}\right) = 0.307$$

$$K_{p_2} = \tan^2\left(45^0 + \frac{\phi_2}{2}\right) = \tan^2\left(45^0 + \frac{32^0}{2}\right) = 3.254,\quad K_{p_2} - K_{a_2} = 2.948$$

Forces per unit length of wall

$$P_1 = 0.5K_{a_1}\gamma_1 d^2 = 0.5\times0.333\times20\times1.2^2 = 4.8\ \text{kN}$$

$$P_2 = 0.5K_{a_1}\gamma_1(H+d)(H-d) = 0.5\times0.333\times20\times(15+1.2)(15-1.2) = 744.4\ \text{kN}$$

$$d_2 = \frac{(H-d)(2H+d)}{3(H+d)} = \frac{(15-1.2)(2\times15+1.2)}{3(15+1.2)} = 8.86\ \text{m}$$

$$P_3 = 0.5K_{a_2}\gamma_1 H z_1 = 0.5\times0.307\times20\times15\times1.74 = 80.13\ \text{kN},\quad z_1 = \frac{K_{a_2}\gamma_1 H}{(K_{p_2}-K_{a_2})\gamma_2} = \frac{0.307\times20\times15}{2.948\times18} = 1.74\ \text{m}$$

For $\phi_2 = 32^0$: $x = 0.059H = 0.059\times15 = 0.885$

$$\sum M_T = 0,\quad R(H-d+x) + P_1\frac{d}{3} - P_2 d_2 - P_3\left(H-d+\frac{z_1}{3}\right) = 0$$

$$R(15-1.2+0.885) + 4.8\times\frac{1.2}{3} - 744.4\times8.86 - 80.13\left(15-1.2+\frac{1.74}{3}\right) = 0,\quad R = 527.46\ \text{kN}$$

$$T = (P_1+P_2+P_3) - R = 4.8+744.4+80.13-527.46 = 301.87\ \text{kN}$$

$$D_0 = z_1 + \sqrt{\frac{6R}{(K_{p_2}-K_{a_2})\gamma_{2_1}}} = 1.74 + \sqrt{\frac{6\times301.87}{2.948\times18}} = 7.58\ \text{m},\quad (\text{assumed } x = z_1)$$

$$D = 1.2D_0 = 1.2\times7.58 = 9.1\ \text{m}$$

$$z_2 = \sqrt{\frac{P_1+P_2+P_3-T}{0.5(K_{p_2}-K_{a_2})\gamma_2}} = \sqrt{\frac{4.8+744.4+80.13-301.87}{0.5\times2.948\times18}} = 4.46\ \text{m}$$

$$M_{max} = (P_1+P_2)\left(\frac{H}{3}+z_1+z_2\right) + P_3\left(\frac{2}{3}z_1+z_2\right) - T(H-d+z_1+z_2) - 0.5(K_{p_2}-K_{a_2})\gamma_2 z_2^2\left(\frac{z_2}{3}\right)$$

$$= (4.8+744.4)\left(\frac{15}{3}+1.74+4.46\right) + 80.13\left(\frac{2}{3}\times1.74+4.46\right) - 301.87(15-1.2+1.74+4.46)$$

$$-0.5\times2.948\times18\times\frac{4.46^3}{3} = 2019.4\ \text{kN}\cdot\text{m/m}$$

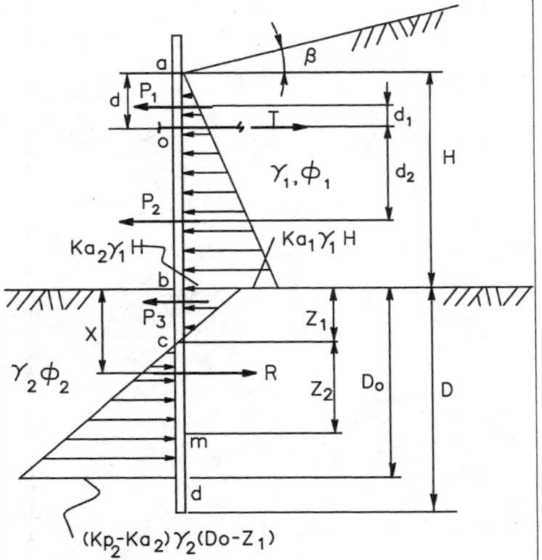

Earth pressure: $P_1 = 0.5K_{a_1}\gamma_1 d^2$, $P_2 = 0.5K_{a_1}\gamma_1 H^2$, $P_3 = 0.5K_{a_2}\gamma_1 z_1^2$

$$d_1 = \frac{d}{3},\quad d_2 = \frac{(H-d)(2H+d)}{3(H+d)},\quad z_1 = \frac{K_{a_2}\gamma_1 H}{(K_{p_2}-K_{a_2})\gamma_2}$$

x = distance to contraflexure point

ϕ	20^0	25^0	30^0	35^0	40^0
x	0.25H	0.15H	0.075H	0.035H	0.007H

May accept $x = z_1$

Equation to determine R: $\sum M_T = 0$

$$R(H-d+x)+P_1\frac{d}{3}-P_2 d_2 - P_3\left(H-d+\frac{z_1}{3}\right)=0$$

T = tension in the anchor rod, $T = (P_1+P_2+P_3)-R$

$$D_0 = z_1 + \sqrt{\frac{6R}{(K_{p_2}-K_{a_2})\gamma_2}},\quad \text{(assumed } x = z_1)$$

$D = (1.2 \text{ to } 1.4)D_0$ for factor of safety at 1.5 to 2.0

m = point of zero shear and maximum bending moment

$$z_2 = \sqrt{\frac{P_1+P_2+P_3-T}{0.5(K_{p_2}-K_{a_2})\gamma_2}}$$

Maximum bending moment :

$$M_{max} = (P_1+P_2)\left(\frac{H}{3}+z_1+z_2\right)+P_3\left(\frac{2z_1}{3}+z_2\right)-T(H-d+z_1+z_2)$$
$$-0.5(K_{p_2}-K_{a_2})\gamma_2 z_2^2\cdot\left(\frac{z_2}{3}\right)$$

NOTES

12, 13. PIPES and TUNNELS

Bending Moments for Various Static Loading Conditions

N O T E S

This chapter provides formulas for computation of bending moments in various structures with rectangular or circular cross-sections, including underground pipes and tunnels. The formulas for structures with circular cross-sections can also be used to compute axial forces and shears.

The formulas provided are applicable to analysis of elastic systems only.

The tables contain the most common cases of loading conditions.

PIPES AND TUNNELS

RECTANGULAR CROSS–SECTION

12.1

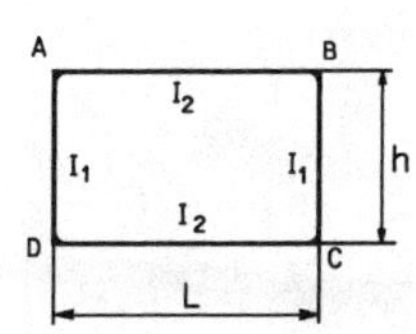

$$k = \frac{I_2 h}{I_1 L}$$

$+M$ = tension on inside of section

1

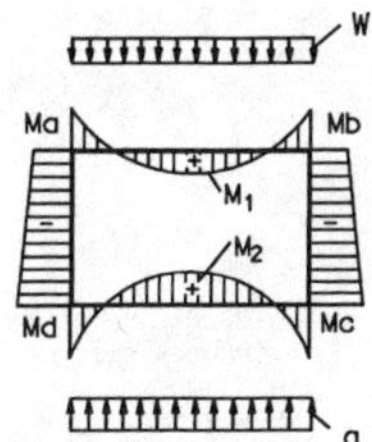

For $q \neq w$

$$M_a = M_b = -\frac{L^2}{12} \cdot \frac{w(2k+3) - qk}{k^2 + 4k + 3}$$

$$M_c = M_d = -\frac{L^2}{12} \cdot \frac{q(2k+3) - wk}{k^2 + 4k + 3}$$

For $q = w$

$$M_a = M_b = M_c = M_d = -\frac{wL^2}{12} \cdot \frac{k+3}{K^2 + 4k + 3}$$

2

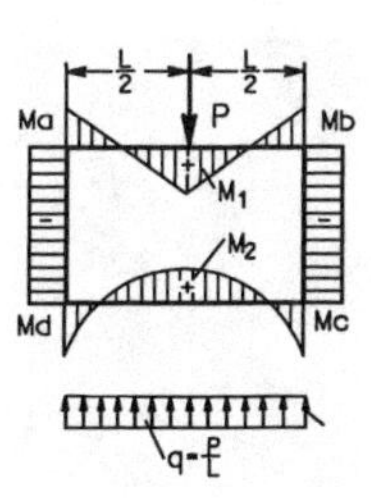

$$M_a = M_b = -\frac{PL}{24} \cdot \frac{4k+9}{k^2 + 4k + 3}$$

$$M_c = M_d = -\frac{PL}{24} \cdot \frac{4k+6}{k^2 + 4k + 3}$$

For $k = 1$

$$M_a = M_b = -\frac{13}{192} PL$$

$$M_c = M_d = -\frac{7}{192} PL$$

NOTES

3

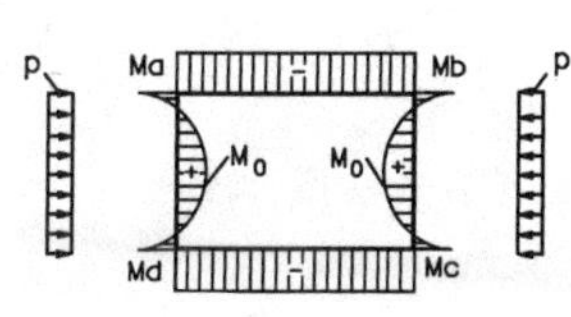

$$M_a = -\frac{ph^2k}{12(k+1)}$$

$$M_b = M_c = M_d = M_a$$

For $k = 1$ and $h = L$

$$M_a = M_b = M_c = M_d = -\frac{ph^2}{24}$$

$$M_0 = 0.125ph^2 - 0.5(M_a + M_d)$$

4

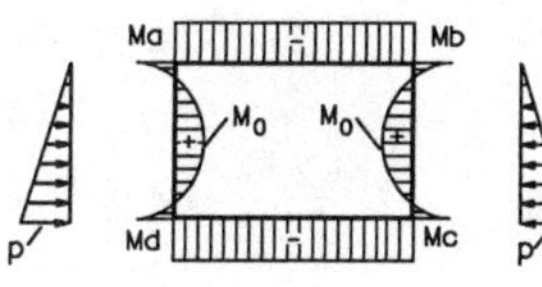

$$M_a = M_b = -\frac{ph^2k(2k+7)}{60(k^2+4k+3)}$$

$$M_c = M_d = -\frac{ph^2k(3k+8)}{60(k^2+4k+3)}$$

For $k = 1$ and $h = L$

$$M_a = M_b = -\frac{3ph^2}{160}, \quad M_c = M_d = -\frac{11ph^2}{480}$$

$$M_0 = 0.064ph^2 - \left[M_a + 0.577(M_d - M_a)\right]$$

5

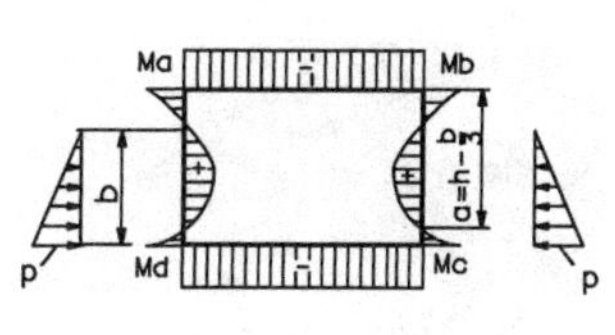

$$M_a = M_b = -\frac{(A+D)(2k+3) - D(3k+3)}{3(k^2+4k+3)}$$

$$M_c = M_d = -\frac{D(3k+3) - (A+D)k}{3(k^2+4k+3)}$$

$$A = \frac{pb^2k}{60h^2}(10h^2 - 3b^2)$$

$$D = \frac{pbak}{2h^2}\left(h^2 - a^2 - b^2\frac{45a - 2b}{270a}\right)$$

N O T E S

Table 12.3

Example. Rectangular pipe 7 in Table 12.3

Given. Concrete frame, $L = 4\text{ m}$, $H = 2.5\text{ m}$, $h_1 = 10\text{ cm}$, $h_2 = 20\text{ cm}$

$b = 1\text{ m}$ (unit length of pipe)

$$I_1 = \frac{bh_1^3}{12} = \frac{100\times10^3}{12} = 8333\text{ cm}^4,\quad I_2 = \frac{bh_2^3}{12} = \frac{100\times20^3}{12} = 66667\text{ cm}^4$$

Uniformly distributed load $w = 120\text{ kN/m}$

Required. Compute bending moments

Solution. $k = \dfrac{I_2H}{I_1L} = \dfrac{66667\times2.5}{8333\times4} = 5.0,\quad r = 2k+1 = 2\times5+1 = 11$

$$m = 20(k+2)\ m = 20(k+2)(6k^2+6k+1) = 20(5+2)(6\times5^2+6\times5+1) = 25340$$

$$\alpha_1 = 138k^2+265k+43 = 138\times5^2+265\times5+43 = 4818$$

$$\alpha_2 = 78k^2+205k+33 = 78\times5^2+205\times5+33 = 3008$$

$$\alpha_3 = 81k^2+148k+37 = 81\times5^2+148\times5+37 = 2802$$

$$\alpha_4 = 21k^2+88k+27 = 21\times5^2+88\times5+27 = 992$$

$$M_a = -\frac{wL^2}{24}\left(\frac{1}{r}+\frac{\alpha_1}{m}\right) = \frac{120\times4^2}{24}\left(\frac{1}{11}+\frac{4818}{25340}\right) = -22.56\text{ kN}\cdot\text{m},\quad M_e = -\frac{wL^2}{24}\left(\frac{1}{r}-\frac{\alpha_1}{m}\right) = +7.92\text{ kN}\cdot\text{m}$$

$$M_c = -\frac{wL^2}{24}\left(\frac{1}{r}+\frac{\alpha_2}{m}\right) = \frac{120\times4^2}{24}\left(\frac{1}{11}+\frac{3008}{25340}\right) = -16.78\text{ kN}\cdot\text{m},\quad M_f = -\frac{wL^2}{24}\left(\frac{1}{r}-\frac{\alpha_2}{m}\right) = +2.24\text{ kN}\cdot\text{m}$$

$$M_{b1} = -\frac{wL^2}{24}\left(\frac{3k+1}{r}+\frac{\alpha_3}{m}\right) = \frac{120\times4^2}{24}\left(\frac{3\times5}{11}+\frac{2802}{25340}\right) = -125.2\text{ kN}\cdot\text{m}$$

$$M_{b2} = -\frac{wL^2}{24}\left(\frac{3k+1}{r}-\frac{\alpha_3}{m}\right) = -107.44\text{ kN}\cdot\text{m},\quad M_{b4} = -\frac{wL^2}{12}\cdot\frac{\alpha_3}{m} = -17.76\text{ kN}\cdot\text{m}$$

$$M_{d4} = -\frac{wL^2}{12}\cdot\frac{\alpha_4}{m} = -\frac{120\times4^2}{12}\cdot\frac{992}{25340} = -6.24\text{ kN}\cdot\text{m}$$

$$M_{d6} = -\frac{wL^2}{24}\left(\frac{3k+1}{r}-\frac{\alpha_4}{m}\right) = -\frac{120\times4^2}{24}\left(\frac{3\times5+1}{11}+\frac{992}{25340}\right) = -119.44\text{ kN}\cdot\text{m}$$

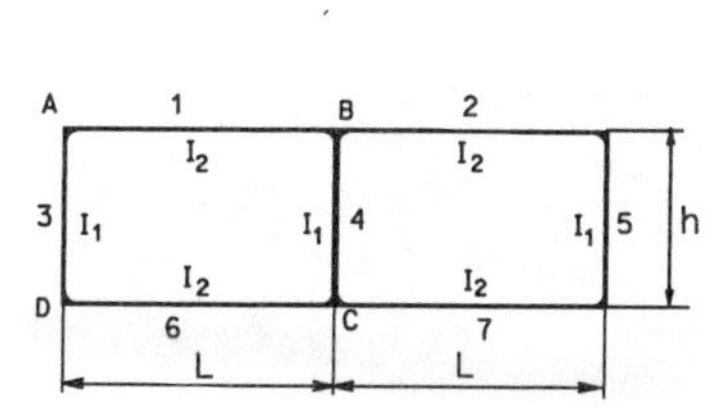

$$k = \frac{I_2 h}{I_1 L}$$

$$r = 2k + 1$$

$$m = 20(k+2)(6k^2 + 6k + 1)$$

$+M$ = tension on inside of section

6

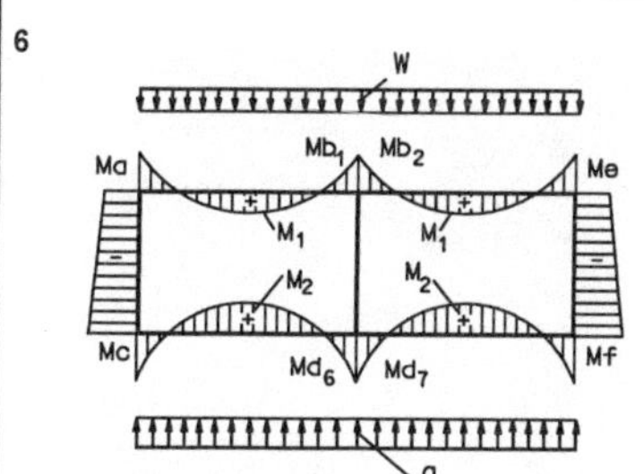

$$q = w$$

$$M_a = -\frac{wL^2}{12} \cdot \frac{1}{r}, \quad M_c = M_e = M_f = M_a$$

$$M_{b1} = -\frac{wL^2}{12} \cdot \frac{3k+1}{r}, \quad M_{b2} = M_{d6} = M_{d7} = M_{b1}$$

$$M_{b4} = M_{d4} = 0$$

7

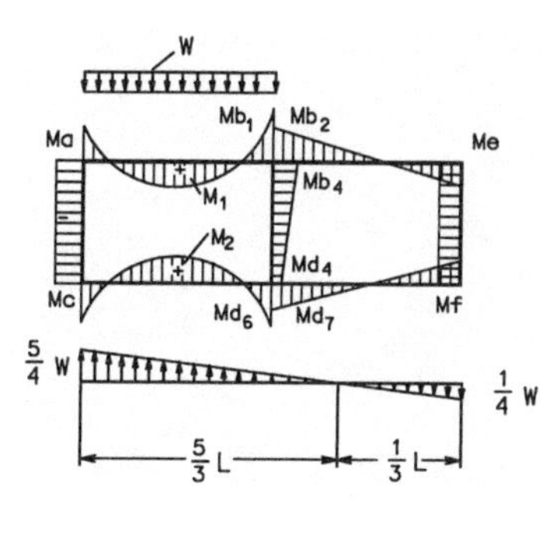

$$M_a = -\frac{wL^2}{24}\left(\frac{1}{r} + \frac{\alpha_1}{m}\right), \quad M_e = -\frac{wL^2}{24}\left(\frac{1}{r} - \frac{\alpha_1}{m}\right)$$

$$M_c = -\frac{wL^2}{24}\left(\frac{1}{r} + \frac{\alpha_2}{m}\right), \quad M_f = -\frac{wL^2}{24}\left(\frac{1}{r} - \frac{\alpha_2}{m}\right)$$

$$M_{b1} = -\frac{wL^2}{24}\left(\frac{3k+1}{r} + \frac{\alpha_3}{m}\right), \quad M_{b4} = -\frac{wL^2}{12} \cdot \frac{\alpha_3}{m}$$

$$M_{b2} = -\frac{wL^2}{24}\left(\frac{3k+1}{r} - \frac{\alpha_3}{m}\right), \quad M_{d4} = -\frac{wL^2}{12} \cdot \frac{\alpha_4}{m}$$

$$M_{d6} = -\frac{wL^2}{24}\left(\frac{3k+1}{r} + \frac{\alpha_4}{m}\right)$$

$$M_{d7} = -\frac{wL^2}{24}\left(\frac{3k+1}{r} - \frac{\alpha_4}{m}\right)$$

$$\alpha_1 = 138k^2 + 265k + 43, \quad \alpha_3 = 81k^2 + 148k + 37$$

$$\alpha_2 = 78k^2 + 205k + 33, \quad \alpha_4 = 21k^2 + 88k + 27$$

NOTES

8

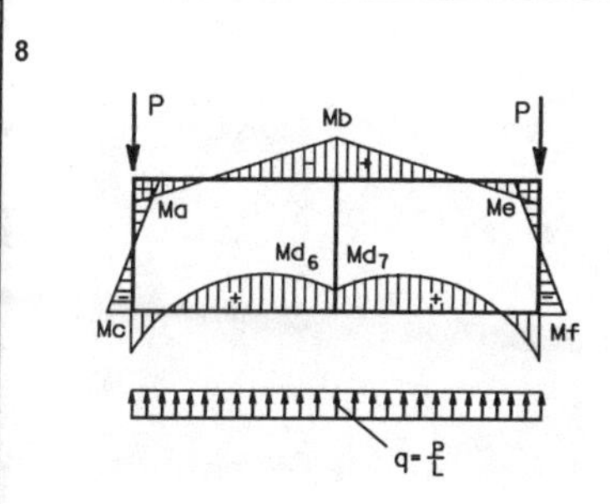

$$m_1 = 24(k+6)r$$

$$M_a = M_e = PL\frac{47k+18}{m_1}$$

$$M_{b1} = M_{b2} = -PL\frac{15k^2+49k+18}{m_1}$$

$$M_c = M_f = -PL\frac{49k+30}{m_1}$$

$$M_{d6} = M_{d7} = PL\frac{9k^2+11k+6}{m_1}$$

$$M_{b4} = M_{d4} = 0$$

9

$$M_a = M_c = M_e = M_f = -\frac{ph^2}{6}\cdot\frac{k}{r}$$

$$M_{b1} = M_{b2} = M_{d7} = \frac{ph^2}{12}\cdot\frac{k}{r}$$

$$M_{b4} = M_{d4} = 0$$

10

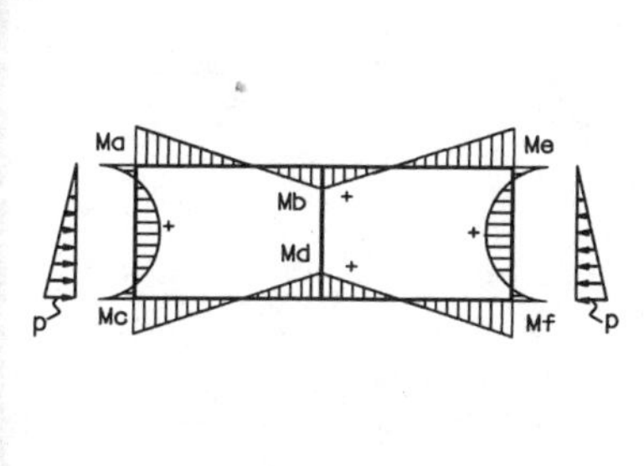

$$m_2 = \frac{20(k+6)r}{k}$$

$$M_a = M_e = -\frac{ph^2}{6}\cdot\frac{8k+59}{m_2}$$

$$M_c = M_f = -\frac{ph^2}{6}\cdot\frac{12k+61}{m_2}$$

$$M_{b1} = M_{b2} = \frac{ph^2}{6}\cdot\frac{7k+31}{m_2}$$

$$M_{d6} = M_{d7} = \frac{ph^2}{6}\cdot\frac{3k+29}{m_2}$$

$$M_{b4} = M_{d4} = 0$$

NOTES

11

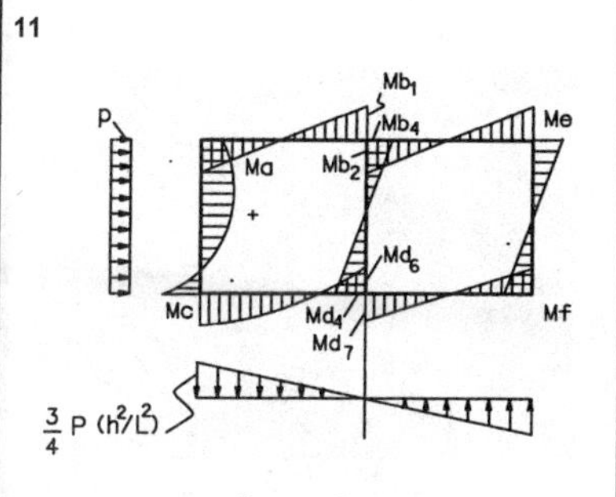

$$\alpha_1 = 120k^3 + 278k^2 + 335k + 63$$
$$\alpha_2 = 360k^3 + 742k^2 + 285k + 27$$
$$\alpha_3 = 120k^3 + 529k^2 + 382k + 63$$
$$\alpha_4 = 120k^3 + 611k^2 + 558k + 87$$

$$m = 20(k+2)(6k^2+6k+1), \quad n_1 = \frac{r}{k}$$

$$M_a = \frac{ph^2}{24}\left(-\frac{2}{n_1} + \frac{\alpha_1}{m}\right), \quad M_e = \frac{ph^2}{24}\left(-\frac{2}{n_1} - \frac{\alpha_1}{m}\right)$$

$$M_c = -\frac{ph^2}{24}\left(\frac{2}{n_1} + \frac{\alpha_2}{m}\right), \quad M_f = -\frac{ph^2}{24}\left(\frac{2}{n_1} - \frac{\alpha_2}{m}\right)$$

$$M_{b1} = -\frac{ph^2}{24}\left(-\frac{1}{n_1} + \frac{\alpha_3}{m}\right), \quad M_{b2} = -\frac{ph^2}{24}\left(-\frac{1}{n_1} - \frac{\alpha_3}{m}\right)$$

$$M_{d6} = -\frac{ph^2}{24}\left(-\frac{1}{n_1} + \frac{\alpha_4}{m}\right), M_{d7} = -\frac{ph^2}{24}\left(-\frac{1}{n_1} - \frac{\alpha_4}{m}\right)$$

$$M_{b4} = -\frac{ph^2}{12}\cdot\frac{\alpha_3}{m}, \quad M_{d4} = \frac{ph^2}{12}\cdot\frac{\alpha_4}{m}$$

12

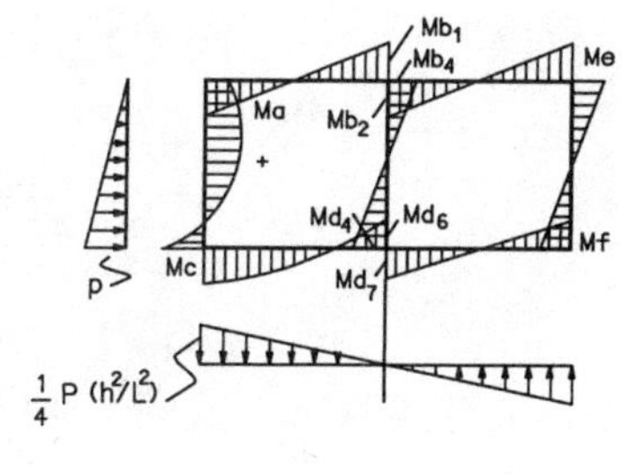

$$\alpha_1 = 24k^3 + 50k^2 + 99k + 21$$
$$\alpha_2 = 144k^3 + 298k^2 + 109k + 9$$
$$\alpha_3 = 36k^3 + 169k^2 + 120k + 21$$
$$\alpha_4 = 36k^3 + 203k^2 + 192k + 29$$

$$m = 20(k+2)(6k^2+6k+1), \quad n_2 = \frac{10(k+6)r}{k}$$

$$\frac{M_a}{M_e} = \frac{ph^2}{24}\left(-\frac{8k+59}{n_2} \pm \frac{\alpha_1}{m}\right)$$

$$\frac{M_c}{M_f} = -\frac{ph^2}{24}\left(\frac{12k+61}{n_2} \pm \frac{\alpha_2}{m}\right)$$

$$\frac{M_{b1}}{M_{b2}} = -\frac{ph^2}{24}\left(-\frac{7k+31}{n_2} \pm \frac{\alpha_3}{m}\right)$$

$$\frac{M_{b1}}{M_{b2}} = -\frac{ph^2}{24}\left(-\frac{7k+31}{n_2} \pm \frac{\alpha_3}{m}\right)$$

$$\frac{M_{d6}}{M_{d7}} = \frac{ph^2}{24}\left(\frac{3k+29}{n_2} \pm \frac{\alpha_4}{m}\right)$$

$$M_{b4} = -\frac{ph^2}{12}\cdot\frac{\alpha_3}{m}, \quad M_{d4} = \frac{ph^2}{12}\cdot\frac{\alpha_4}{m}$$

NOTES

PIPES AND TUNNELS

CIRCULAR CROSS-SECTION

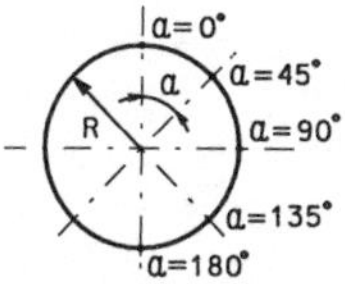

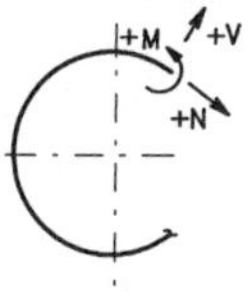

13.1

+M = tension on inside of ring

+ Tension

− Compression

Loading condition		$\alpha = 0$	$\alpha = 45^0$	$\alpha = 90^0$	$\alpha = 135^0$	$\alpha = 180^0$
1 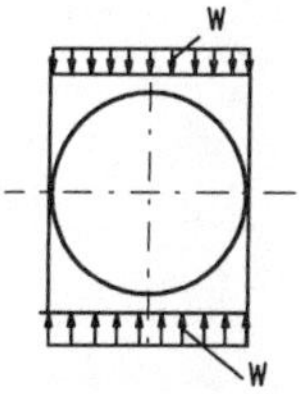	M	$+0.25wR^2$	0	$-0.25wR^2$	0	$+0.25wR^2$
	N	0	$-0.5wR$	$-1.0wR$	$-0.5wR$	0
	V	0	$-0.5wR$	0	$+0.5wR$	0
2 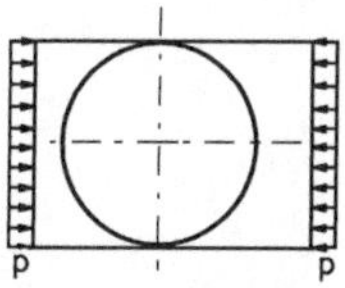	M	$-0.25pR^2$	0	$+0.25pR^2$	0	$-0.25pR^2$
	N	$-1.0pR$	$-0.5pR$	0	$-0.5pR$	$-1.0pR$
	V	0	$+0.5pR$	0	$-0.5pR$	0
3 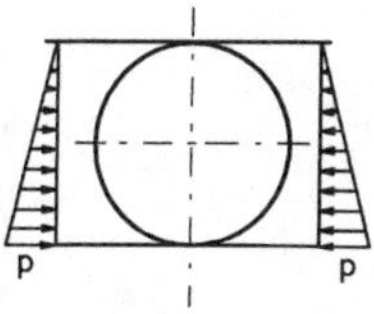	M	$-0.208pR^3$	$-0.029pR^3$	$+0.25pR^3$	$+0.029pR^3$	$-0.292pR^3$
	N	$-0.625pR^2$	$-0.412pR^2$	0	$-0.588pR^2$	$-1.375pR^2$
	V	0	$+0.411pR^2$	$+0.125pR^2$	$-0.589pR^2$	0
4 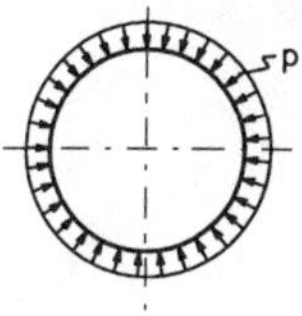	M	0	0	0	0	0
	N	$-pR$	$-pR$	$-pR$	$-pR$	$-pR$
	V	0	0	0	0	0

NOTES

CIRCULAR CROSS-SECTION

Loading condition		$\alpha = 0$	$\alpha = 45^0$	$\alpha = 90^0$	$\alpha = 135^0$	$\alpha = 180^0$
5 (Rγ; 0.2145 γR)	M	$+0.027\gamma R^3$	$+0.010\gamma R^3$	$-0.042\gamma R^3$	$-0.003\gamma R^3$	$+0.045\gamma R^3$
	N	$+0.021\gamma R^2$	$-0.030\gamma R^2$	$-0.215\gamma R^2$	$-0.122\gamma R^2$	$-0.021\gamma R^2$
	V	0	$-0.061\gamma R^2$	$-0.021\gamma R^2$	$+0.092\gamma R^2$	0
6 Buoyancy Forces ($\gamma_w R(1-\cos\alpha)$)	M	0	0	0	0	0
	N	$-0.5\gamma_w R^2$	$-0.646\gamma_w R^2$	$-1.0\gamma_w R^2$	$-1.354\gamma_w R^2$	$-1.5\gamma_w R^2$
	V	0	0	0	0	0
7 (45°; 30°, 30°; F, F)	M	$+0.151\gamma_w R^3$	$+0.026\gamma_w R^3$	$-0.176\gamma_w R^3$	$+0.001\gamma_w R^3$	$+0.121\gamma_w R^3$
	N	$-0.481\gamma_w R^2$	$+0.188\gamma_w R^2$	$+0.066\gamma_w R^2$	$+0.316\gamma_w R^2$	$+1.077\gamma_w R^2$
	V	0	$+0.191\gamma_w R^2$	$+0.016\gamma_w R^2$	$-0.567\gamma_w R^2$	0
8 (30°, 30°; F, F)	M	$+0.320\gamma_w R^3$	$+0.152\gamma_w R^3$	$-0.091\gamma_w R^3$	$+0.128\gamma_w R^3$	$+0.279\gamma_w R^3$
	N	$-0.821\gamma_w R^2$	$-0.653\gamma_w R^2$	$+0.090\gamma_w R^2$	$+1.366\gamma_w R^2$	$+1.5\gamma_w R^2$
	V	0	$+0.366\gamma_w R^2$	$+0.125\gamma_w R^2$	$-0.744\gamma_w R^2$	0

γ and γ_w = unit weight of soil and liquid, respectively

NOTES

Loading condition		$\alpha=0$	$\alpha=45^0$	$\alpha=90^0$	$\alpha=135^0$	$\alpha=180^0$
9 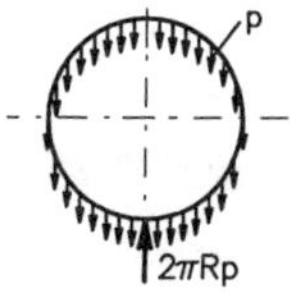	M	$+0.378pR^2$	$+0.043pR^2$	$-0.442pR^2$	$-0.007pR^2$	$+0.308pR^2$
	N	+0.25pR	−0.378pR	−1.570pR	−1.842pR	−0.25pR
	V	0	−0.732pR	+0.25pR	−1.488pR	0
10 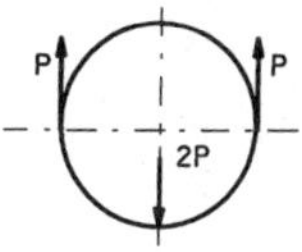	M	−0.137PR	−0.043PR	+0.182PR	+0.114PR	−0.500PR
	N	−0.318P	−0.225P	+1.0P	+0.939P	+0.318P
	V	0	−0.225P	−0.318P	+0.482P	+1.0P
11 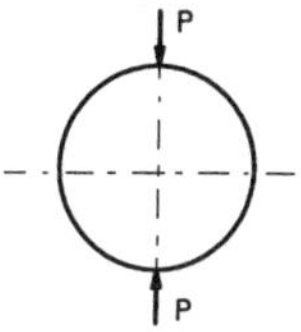	M	+0.318PR	+0.035PR	−0.182PR	+0.035PR	+0.318PR
	N	0	−0.354P	−0.5P	−0.354P	0
	V	+0.5P	+0.354P	0	−0.354P	−0.5P
12 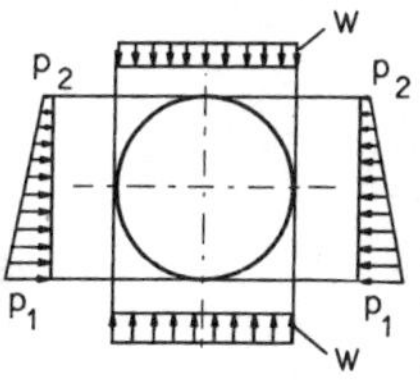	$M_{max}=\frac{wR^2}{4}-\frac{R^2}{48}(5p_1+7p_2)$					
	$M_{min}=-\frac{wR^2}{4}+\frac{R^2}{8}(p_1+p_2)$					
	$N=\frac{R(11p_1+5p_2)}{16}$					
	If $p_1=p_2=p$: $M_{max}=\frac{R^2}{4}(w-p)$, $M_{min}=-\frac{R^2}{4}(w-p)$					
	$N=pR$					

NOTES

APPENDIX

NOTES

UNITS

CONVERSION between ANGLO-AMERICAN and METRIC SYSTEMS — U.1

Metric Units	Conversion Factors	
Units of Length		
millimeter (mm)	1 inch$(\text{in}) = 25.4(\text{mm})$	$1(\text{mm}) = 0.03937(\text{in})$
1 centimeter$(\text{cm}) = 10(\text{mm})$	1 foot$(\text{ft}) = 12(\text{in}) = 304.8(\text{mm})$	$1(\text{cm}) = 0.3937(\text{in})$
1 decimeter$(\text{dm}) = 10(\text{cm}) = 100(\text{mm})$	1 yard$(\text{yd}) = 3(\text{ft}) = 0.9144(\text{m})$	$1(\text{m}) = 1.0904(\text{yd})$
1 meter$(\text{m}) = 100(\text{cm}) = 1000(\text{mm})$	1 mile $= 1760(\text{yd}) = 1609.344(\text{m})$	$1(\text{km}) = 3281(\text{ft})$
1 kilometer$(\text{km}) = 1000(\text{m})$	1 mile $= 1.6093(\text{km})$	$1(\text{km}) = 0.6214$ mile
Units of Area		
square millimeter (mm^2)	1 square inch$(\text{in}^2) = 645.16(\text{mm}^2)$	$1(\text{mm}^2) = 0.001550(\text{in}^2)$
1 square centimeter $(\text{cm}^2) = 100(\text{mm}^2)$	1 square foot$(\text{ft}^2) = 0.092903(\text{m}^2)$	$1(\text{cm}^2) = 0.1550(\text{in}^2)$
1 square meter$(\text{m}^2) = 10^6(\text{mm}^2)$	1 square yard$(\text{yd}^2) = 0.836127(\text{m}^2)$	$1(\text{m}^2) = 10.76(\text{ft}^2)$
1 square kilometer$(\text{km}^2) = 10^6(\text{m}^2)$	1 acre $= 4046.856(\text{m}^2)$	$1(\text{m}^2) = 1.19599(\text{yd}^2)$
1 hectare$(\text{ha}) = 10^4(\text{m}^2) = 0.01(\text{km}^2)$	1 square mile $= 2.5898(\text{km}^2)$	$1(\text{km}^2) = 0.3861$ square mile
Units of Volume		
cubic millimeter (mm^3)	1 cubic inch$(\text{in}^3) = 16387.064(\text{mm}^3)$	$1(\text{mm}^3) = 0.00006102(\text{in}^3)$
1 cubic centimeter$(\text{cm}^3) = 10^3(\text{mm}^3)$	1 cubic foot$(\text{ft}^3) = 0.02831685(\text{m}^3)$	$1(\text{cm}^3) = 0.06102(\text{in}^3)$
1 cubic meter$(\text{m}^3) = 10^9(\text{mm}^3)$	1 cubic yard$(\text{yd}^3) = 0.764555(\text{m}^3)$	$1(\text{m}^3) = 1.30795(\text{yd}^3)$
1 cubic kilometer$(\text{km}^3) = 10^9(\text{m}^3)$	1 acre · foot $= 1233.482(\text{m}^3)$	$1(\text{m}^3) = 35.31(\text{ft}^3)$
1 liter$(\text{L}) = 1000(\text{cm}^3) = 0.001(\text{m}^3)$	1 gallon $= 3.785412$ liters(L)	$1(\text{L}) = 0.264172$ gallon

NOTES

UNITS

CONVERSION between ANGLO-AMERICAN and METRIC SYSTEMS — U.2

Metric Units	Conversion Factors	
Units of Mass		
milligram(mg) 1 gram$(g) = 1000(mg)$ 1 kilogram$(kg) = 1000(g)$ 1 ton$(t) = 1000(kg)$	1 ounce $= 28.34952(g)$ 1 pound$(lb) = 0.453592(kg)$ 1 kip $= 453.592(kg)$ 1 ton$(2000\ lb) = 907.184(kg)$	Mass per unit length $1\ (lb/ft) = 1.48816(kg/m)$ Mass per unit area $1(lb/ft^2) = 4.88243(kg/m^2)$ Mass per unit volume (mass density) $1(lb/ft^3) = 16.01846(kg/m^3)$ $1(lb/yd^3) = 0.593276(kg/m^3)$
Units of Force		
1 newton$(N) = 1\ kg(mass)/(m/sec^2)$ 1 kilonewton$(kN) = 1000(N)$ 1 meganewton$(MN) = 1000(kN)$ Gravitational force: $1(N) = 1\ kg(mass)/9.81 = 0.102(kg)$ or 1 kg$(force) = 9.81(N)$ Unit weight: $1(lb/ft^3) = 0.1571(kN/m^3)$	$1(lb) = 4.448222(N)$ 1 kip $= 4.448222(kN)$ 1 ton$(2000\ lb) = 8.896444(kN)$ $1(N) = 0.2248(lb)$ $1(kN) = 0.2248$ kip $1(kN) = 0.1124$ ton $1(kN/m^3) = 6.366(lb/ft^3)$	Force per unit length $1(lb/in) = 175.1268(N/m)$ $1(lb/ft) = 14.5939(N/m)$ Moment of force $1(lb \cdot in) = 0.112985(N \cdot m)$ $1(lb \cdot ft) = 1.355818(N \cdot m)$
Units of Pressure, Stress, Modulus of Elasticity		
1 pascal$(Pa) = 1(N/m^2)$ 1 kilopascal$(kPa) = 1000(Pa) =$ 1 kN/m² 1 megapascal$(MPa) = 1000(kPa)$ 1 gigapascal$(GPa) = 1000(MPa)$ 1 atmosphere$(atm) = 1(kg/cm^2) = 98.1(kPa)$ 1 bar $= 1.02(kg/cm^2) = 100(kPa)$	$1(lb/in^2) = 6.894757(kPa)$ $1(kip/in^2) = 6.894757(MPa)$ $1(lb/ft^2) = 47.88026(Pa)$ $1(kip/ft^2) = 47.88026(kPa)$ $1(lb/in^2) = 0.07029(kg/cm^2)$	$1(kPa) = 0.145038(lb/in^2)$ $1(MPa) = 0.145038(kip/in^2)$ $1(Pa) = 0.020885(lb/ft^2)$ $1(kPa) = 0.020885(kip/ft^2)$ $1(kg/cm^2) = 14.23(lb/in^2)$

Temperature: $T_C^0 = \frac{5}{9}(T_F^0 - 32^0)$, where T_C^0 and T_F^0 are Celsius and Fahrenheit temperatures, respectively.

NOTES

MATHEMATICAL FORMULAS

ALGEBRA

M.1

POWERS		ROOTS	
$a^m \cdot a^n = a^{m+n}$	$\frac{a^m}{a^n} = a^{m-n}$	$a^{\frac{m}{n}} = \sqrt[n]{a^m}$	$\sqrt[m \cdot n]{a^m} = \sqrt[n]{a}$
$\left(a^m\right)^n = a^{m \cdot n}$	$(a \cdot b)^m = a^m \cdot b^m$	$\sqrt[m]{a} \cdot \sqrt[m]{b} = \sqrt[m]{a \cdot b}$	$\frac{\sqrt[m]{a}}{\sqrt[m]{b}} = \sqrt[m]{\frac{a}{b}}$
$\left(\frac{a}{b}\right)^m = \frac{a^m}{b^m}$	$a^m \cdot b \pm a^m \cdot c = (b \pm c) a^m$	$\left(\sqrt[m]{a}\right)^n = \sqrt[m]{a^n}$	$\sqrt[m]{\sqrt[n]{a}} = \sqrt[m \cdot n]{a}$
$a^{-m} = \frac{1}{a^m}$	$a^0 = 1$, when $a \neq 0$	$i = \sqrt{-1}$	$\sqrt{-a} = i \cdot \sqrt{a}$

LOGARITHMS	$\log_a N = n$ a = base, N = antilogarithm, n = logarithm (log) $\log_{10} = \lg$ = common log, $\log_e = \ln$ = natural log
$\log_a (x \cdot y) = \log_a x + \log_a y$	$e = 2.718281828459...$
$\log_a \left(\frac{x}{y}\right) = \log_a x - \log_a y$	$\lg 0.01 = -2, \ \lg 0.1 = -1, \ \lg 1 = 0,$ $\lg 10 = 1, \ \lg 100 = 2$
$\log_a x^m = m \cdot \log_a x$	$\lg x = \lg e \cdot \ln x = 0.434294 \cdot \ln x$
$\log_a \sqrt[m]{x} = \frac{1}{m} \log_a x$	$\ln x = \frac{\lg x}{\lg e} = 2.302585 \cdot \lg x$

FACTORIAL	
	$n! = 1 \cdot 2 \cdot 3 \cdot ... \cdot n$
	$(n+1)! = (n+1) n!$
	$0! = 1, \quad (0+1)! = (0+1) 0!$
	$n! \approx \sqrt{2\pi n} \left(\frac{n}{e}\right)^n$

PERMUTATIONS	COMBINATIONS
$P_m^n = \frac{n!}{(n-m)!} = n \cdot (n-1) \cdot (n-2) \cdot ... (n-m+1)$ $n \geq m$	$C_m^n = \frac{n!}{m!(n-m)!}$ $n \geq m$
Example: $P_3^5 = \frac{1 \cdot 2 \cdot 3 \cdot 4 \cdot 5}{1 \cdot 2} = 60$	**Example:** $C_3^5 = \frac{1 \cdot 2 \cdot 3 \cdot 4 \cdot 5}{1 \cdot 2 \cdot 3 \cdot (1 \cdot 2)} = 10$

Where : P = number of possible permutations, C = number of possible combinations,
n = number of things given, m = number of selections from n given things.

NOTES

MATHEMATICAL FORMULAS

ALGEBRA

M.2

ALGEBRAIC EXPRESSIONS

$(a \pm b)^2 = a^2 \pm 2ab + b^2$	$a^2 - b^2 = (a+b)(a-b)$
$(a \pm b)^3 = a^3 \pm 3a^2b + 3ab^2 \pm b^3$	$a^3 \pm b^3 = (a \pm b)(a^2 \mp ab + b^2)$

$$(a+b)^n = a^n + \frac{n}{1}a^{n-1}b + \frac{n(n-1)}{1\cdot 2}a^{n-2}b^2 + \frac{n(n-1)(n-2)}{1\cdot 2\cdot 3}a^{n-3}b^3 + \ldots b^n$$

$$a^n - b^n = (a-b)\left(a^{n-1} + a^{n-2}b + a^{n-3}b^2 + \ldots + ab^{n-2} + b^{n-1}\right)$$

ALGEBRAIC EQUATIONS

Linear equations

Third-order determinants:

$$a_{11}x + a_{12}y + a_{13}z = b_1$$
$$a_{21}x + a_{22}y + a_{23}z = b_2$$
$$a_{31}x + a_{32}y + a_{33}z = b_3$$

$$x = \frac{D_1}{D}, \quad y = \frac{D_2}{D}, \quad z = \frac{D_3}{D}$$

$$D = \begin{vmatrix} a_{11} & a_{12} & a_{13} \\ a_{21} & a_{22} & a_{23} \\ a_{31} & a_{32} & a_{33} \end{vmatrix} \begin{aligned} &= a_{11}\cdot a_{22}\cdot a_{33} - a_{11}\cdot a_{23}\cdot a_{32} + \\ &+ a_{12}\cdot a_{23}\cdot a_{31} - a_{12}\cdot a_{21}\cdot a_{33} + \\ &+ a_{13}\cdot a_{21}\cdot a_{32} - a_{13}\cdot a_{22}\cdot a_{31} \end{aligned}$$

$$D_1 = \begin{vmatrix} b_1 & a_{12} & a_{13} \\ b_2 & a_{22} & a_{23} \\ b_3 & a_{32} & a_{33} \end{vmatrix} \begin{aligned} &= b_1\cdot a_{22}\cdot a_{33} - b_1\cdot a_{23}\cdot a_{32} + \\ &+ a_{12}\cdot a_{23}\cdot b_3 - a_{12}\cdot b_2\cdot a_{33} + \\ &+ a_{13}\cdot b_2\cdot a_{32} - a_{13}\cdot a_{22}\cdot b_3 \end{aligned}$$

Determine D_2 and D_3 similarly by replacing the y- and z- columns by the b- column

Equation of the 2nd degree

$$x^2 + px + q = 0 \qquad x_{1,2} = -\frac{p}{2} \pm \sqrt{\left(\frac{p}{2}\right)^2 - q}$$

Equation of the 3rd degree

$$x^3 + ax^2 + bx + c = 0$$

$x_1 = y_1 - \frac{a}{3}$	Determinant: $D = \left(\frac{p}{3}\right)^3 + \left(\frac{q}{2}\right)^2$, $p = b - \frac{a^3}{3}$, $q = \frac{2}{27}a^3 - \frac{1}{3}a\cdot b + c$
$x_2 = y_2 - \frac{a}{3}$	If $D = 0$: $y_1 = \sqrt[3]{-4q}$, $y_2 = y_3 = \sqrt[3]{\frac{q}{2}}$
$x_3 = y_3 - \frac{a}{3}$	If $D>0$: $\omega_1 = \frac{-1+i\sqrt{3}}{2}$, $\omega_2 = \frac{-1-i\sqrt{3}}{2}$

$$y_1 = \sqrt[3]{-\frac{q}{2}+\sqrt{D}} + \sqrt[3]{-\frac{q}{2}-\sqrt{D}}, \quad y_2 = \omega_1\sqrt[3]{-\frac{q}{2}+\sqrt{D}} + \omega_2\sqrt[3]{-\frac{q}{2}-\sqrt{D}}, \quad y_3 = \omega_2\sqrt[3]{-\frac{q}{2}+\sqrt{D}} + \omega_1\sqrt[3]{-\frac{q}{2}-\sqrt{D}}$$

If $D<0$: $y_1 = \frac{2}{3}\sqrt{3}\sqrt{|p|}\cos\varphi$, $\quad y_2 = \frac{2}{3}\sqrt{3}\sqrt{|p|}\cos(\varphi + 120^0)$, $\quad y_3 = \frac{2}{3}\sqrt{3}\sqrt{|p|}\cos(\varphi - 120^0)$

$$\varphi = \frac{1}{3}\arccos\frac{-3\sqrt{3q}}{2\sqrt{p^3}}$$

NOTES

MATHEMATICAL FORMULAS

GEOMETRY

SOLID BODIES

M.3

V = volume,	A = cross - section area,	A_s = surface area,	A_m = generated surface
Cuboid 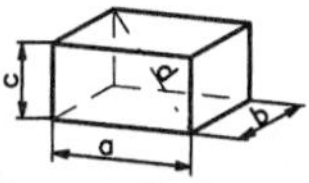	$V = a \cdot b \cdot c$ $A_s = 2(a \cdot b + a \cdot c + b \cdot c)$ $d = \sqrt{a^2 + b^2 + c^2}$	**Cone** 	$V = \frac{\pi}{3} r^2 h$ $A_m = \pi r L,\ \ A_s = \pi r (r + L)$ $L = \sqrt{r^2 + h^2}$
Triangular Prism 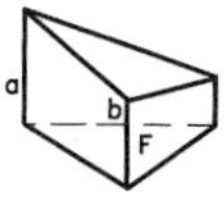	$V = \frac{1}{3}(a + b + c) A$	**Frustum of Cone** 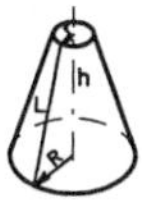	$V = \frac{\pi h}{3}(R^2 + r^2 + Rr)$ $A_m = 2\pi \cdot \rho \cdot L$ $\rho = 0.5(R + r)$ $L = \sqrt{(R^2 - r^2) + h^2}$
Pyramid 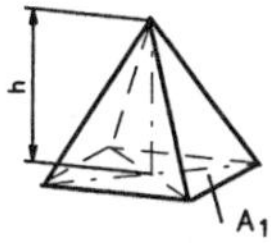	$V = \frac{A_1 h}{3}$	**Sphere** 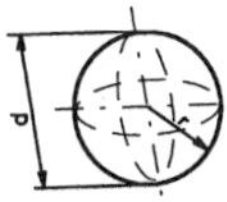	$V = \frac{4}{3} \pi r^3 = 4.189 r^3$ $= \frac{1}{6} \pi d^3 = 0.5236 d^3$ $A_s = 4\pi r^2 = \pi d^2$
Frustum of Pyramid 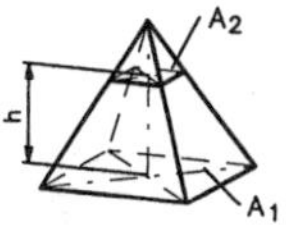	$V = \frac{h}{3}\left(A_1 + A_2 + \sqrt{A_1 A_2}\right)$	**Segment of a Sphere** 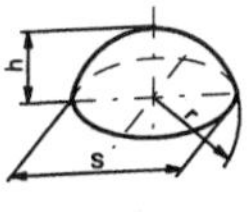	$V = \frac{\pi}{6} h \left(\frac{3}{4} s^2 + h^2\right)$ $= \pi h^2 \left(r - \frac{h}{3}\right)$ $A_m = \frac{\pi}{4}(s^2 + 4h^2) = 2\pi r h$
Cylinder 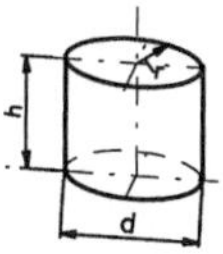	$V = \frac{\pi}{4} d^2 h$ $A_m = 2\pi r h$ $A_s = 2\pi r (r + h)$	**Sector of a Sphere** 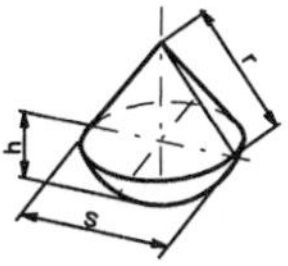	$V = \frac{2}{3} \pi r^2 h$ $A_s = \frac{\pi}{2} r (4h + s)$

NOTES

MATHEMATICAL FORMULAS

GEOMETRY

SOLID BODIES

M.4

Zone of a Sphere

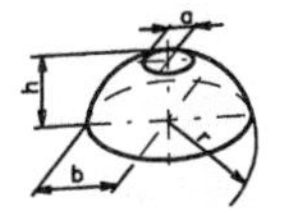

$$V = \frac{\pi}{6} h\left(3a^2 + 3b^2 + h^2\right)$$

$$A_s = \pi\left(2rh + a^2 + b^2\right)$$

$$A_m = 2\pi rh$$

Ungula

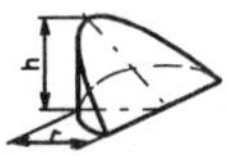

$$V = \frac{2}{3} r^2 h$$

$$A_s = A_m + \frac{\pi}{2}\left(r^2 + r\sqrt{r^2 + h^2}\right)$$

$$A_m = \pi dh$$

Sliced Cylinder

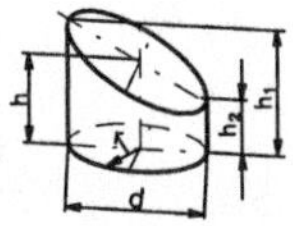

$$V = \frac{\pi}{4} d^2 h$$

$$A_s = \pi r\left[h_1 + h_2 + r + \sqrt{r^2 + \left(h_1 - h_2\right)^2 / 4}\right]$$

$$A_m = \pi dh$$

Barrel

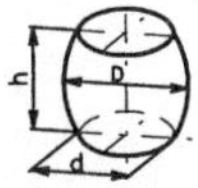

$$V = \frac{\pi}{12} h\left(2D^2 + d^2\right)$$

PLANE ANALYTIC GEOMETRY (Equations)

Straight Line

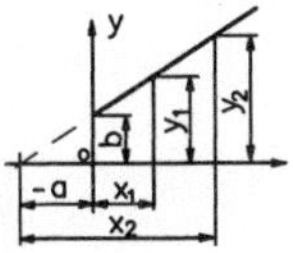

$$y = mx + b$$

$$m = \frac{y_2 - y_1}{x_2 - x_1} = \tan \varphi$$

Circle

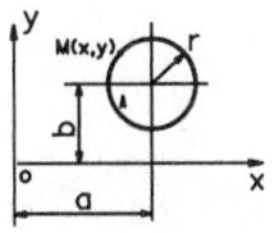

$$\left(x - a\right)^2 + \left(y - b\right)^2 = r^2$$

If $a = 0, \ b = 0$:

$$x^2 + y^2 = r^2$$

Ellipse

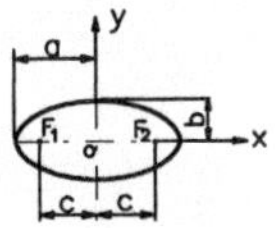

$$\frac{x^2}{a^2} + \frac{y^2}{b^2} = 1$$

$$c = \sqrt{a^2 - b^2}$$

$$\varepsilon = \frac{c}{a} < 1$$

Hyperbola

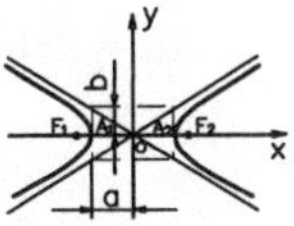

$$\frac{x^2}{a^2} - \frac{y^2}{b^2} = 1$$

$$c = \sqrt{a^2 + b^2}$$

$$\varepsilon = \frac{c}{a} > 1$$

Parabola

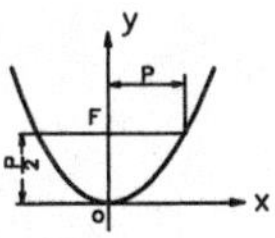

$$x^2 = 2py$$

$$0F = \frac{p}{2}$$

Parabolic Arch

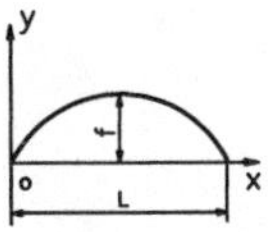

$$y = \frac{4f}{L^2} x\left(L - x\right)$$

N O T E S

TRIGONOMETRY

M.5

BASIC CONVERSIONS			
$\tan\alpha=\dfrac{\sin\alpha}{\cos\alpha}$	$\sec\alpha=\dfrac{1}{\cos\alpha}$	$\sin^2\alpha+\cos^2\alpha=1$	$\dfrac{1}{\cos^2\alpha}=1+\tan^2\alpha$
$\cot\alpha=\dfrac{\cos\alpha}{\sin\alpha}$	$\operatorname{cosec}\alpha=\dfrac{1}{\sin\alpha}$	$\tan\alpha\cdot\cot\alpha=1$	$\dfrac{1}{\sin^2\alpha}=1+\cot^2\alpha$

$\sin(\alpha\pm\beta)=\sin\alpha\cdot\cos\alpha\pm\cos\alpha\cdot\sin\beta$	$\tan(\alpha\pm\beta)=\dfrac{\tan\alpha\pm\tan\beta}{1\mp\tan\alpha\cdot\tan\beta}$
$\cos(\alpha\pm\beta)=\cos\alpha\cdot\cos\beta\mp\sin\alpha\cdot\sin\beta$	$\cot(\alpha\pm\beta)=\dfrac{\cot\alpha\cdot\cot\beta\mp1}{\cot\beta\pm\cot\alpha}$
$\sin2\alpha=2\sin\alpha\cdot\cos\alpha$	$\tan2\alpha=\dfrac{2\tan\alpha}{1-\tan^2\alpha}$
$\cos2\alpha=\cos^2\alpha-\sin^2\alpha$	$\cot2\alpha=\dfrac{\cot^2\alpha-1}{2\cot\alpha}$
$\sin3\alpha=3\sin\alpha-4\sin^3\alpha$	$\tan3\alpha=\dfrac{3\tan\alpha-\tan^3\alpha}{1-3\tan^2\alpha}$
$\cos3\alpha=4\cos^3\alpha-3\cos\alpha$	$\cot3\alpha=\dfrac{\cot^3\alpha-3\cot\alpha}{3\cot^2\alpha-1}$
$\sin\dfrac{\alpha}{2}=\sqrt{\dfrac{1-\cos\alpha}{2}}$	$\tan\dfrac{\alpha}{2}=\dfrac{\sin\alpha}{1+\cos\alpha}=\dfrac{1-\cos\alpha}{\sin\alpha}=\sqrt{\dfrac{1-\cos\alpha}{1+\cos\alpha}}$
$\cos\dfrac{\alpha}{2}=\sqrt{\dfrac{1+\cos\alpha}{2}}$	$\cot\dfrac{\alpha}{2}=\dfrac{\sin\alpha}{1-\cos\alpha}=\dfrac{1+\cos\alpha}{\sin\alpha}=\sqrt{\dfrac{1+\cos\alpha}{1-\cos\alpha}}$
$\sin\alpha=2\sin\dfrac{\alpha}{2}\cdot\cos\dfrac{\alpha}{2}=\dfrac{2\tan\frac{\alpha}{2}}{1+\tan^2\frac{\alpha}{2}}$	$\tan\alpha=\dfrac{2\tan\frac{\alpha}{2}}{1-\tan^2\frac{\alpha}{2}}$
$\cos\alpha=\cos^2\dfrac{\alpha}{2}-\sin^2\dfrac{\alpha}{2}=\dfrac{1-\tan^2\frac{\alpha}{2}}{1+\tan^2\frac{\alpha}{2}}$	$\cot\alpha=\dfrac{\cot^2\frac{\alpha}{2}-1}{2\cot\frac{\alpha}{2}}$
$\sin\alpha+\sin\beta=2\sin\dfrac{\alpha+\beta}{2}\cdot\cos\dfrac{\alpha-\beta}{2}$	$\cos\alpha+\cos\beta=2\cos\dfrac{\alpha+\beta}{2}\cdot\cos\dfrac{\alpha-\beta}{2}$
$\sin\alpha-\sin\beta=2\cos\dfrac{\alpha+\beta}{2}\cdot\sin\dfrac{\alpha-\beta}{2}$	$\cos\alpha-\cos\beta=-2\sin\dfrac{\alpha+\beta}{2}\cdot\sin\dfrac{\alpha-\beta}{2}$
$\tan\alpha\pm\tan\beta=\dfrac{\sin(\alpha\pm\beta)}{\cos\alpha\cdot\cos\beta}$	$\cot\alpha\pm\cot\beta=\dfrac{\sin(\beta\pm\alpha)}{\sin\alpha\cdot\sin\beta}$

NOTES

MATHEMATICAL FORMULAS

TRIGONOMETRY

M.6

BASIC CONVERSIONS	
$\sin\alpha\cdot\cos\beta=\frac{1}{2}\sin(\alpha+\beta)+\frac{1}{2}\sin(\alpha-\beta)$	$\tan\alpha\cdot\tan\beta=\frac{\tan\alpha+\tan\beta}{\cot\alpha+\cot\beta}$
$\cos\alpha\cdot\cos\beta=\frac{1}{2}\cos(\alpha+\beta)+\frac{1}{2}\cos(\alpha-\beta)$	$\cot\alpha\cdot\cot\beta=\frac{\cot\alpha+\cot\beta}{\tan\alpha+\tan\beta}$
$\sin\alpha\cdot\sin\beta=\frac{1}{2}\cos(\alpha-\beta)-\frac{1}{2}\cos(\alpha+\beta)$	$\cot\alpha\cdot\tan\beta=\frac{\cot\alpha+\tan\beta}{\tan\alpha+\cot\beta}$
$\sin^2\alpha-\sin^2\beta=\sin(\alpha+\beta)\cdot\sin(\alpha-\beta)$	$\cos\alpha+\sin\alpha=\sqrt{2}\cdot\sin(45^0+\alpha)$
$\cos^2\alpha-\sin^2\beta=\cos(\alpha+\beta)\cdot\cos(\alpha-\beta)$	$\cos\alpha-\sin\alpha=\sqrt{2}\cdot\cos(45^0+\alpha)$

α^0	0^0	30^0	45^0	60^0	90^0
$\alpha(\text{rad})$	0.0	$\frac{\pi}{6}=0.5236$	$\frac{\pi}{4}=0.7854$	$\frac{\pi}{3}=1.0472$	$\frac{\pi}{2}=1.5708$
$\sin\alpha$	0.0	$\frac{1}{2}=0.5000$	$\frac{\sqrt{2}}{2}=0.7071$	$\frac{\sqrt{3}}{2}=0.8660$	1.0
$\cos\alpha$	1.0	$\frac{\sqrt{3}}{2}=0.8660$	$\frac{\sqrt{2}}{2}=0.7071$	$\frac{1}{2}=0.5000$	0.0
$\tan\alpha$	0.0	$\frac{\sqrt{3}}{3}=0.5774$	1.0	$\sqrt{3}=1.7321$	$\pm\infty$
$\cot\alpha$	$\mp\infty$	$\sqrt{3}=1.7321$	1.0	$\frac{\sqrt{3}}{3}=0.5774$	0.0

φ	$-\alpha$	$90^0\pm\alpha$	$180^0\pm\alpha$	$270^0\pm\alpha$	$360^0-\alpha$
$\sin\varphi$	$-\sin\alpha$	$+\cos\alpha$	$\mp\sin\alpha$	$-\cos\alpha$	$-\sin\alpha$
$\cos\varphi$	$+\cos\alpha$	$\mp\sin\alpha$	$-\cos\alpha$	$\pm\sin\alpha$	$+\cos\alpha$
$\tan\varphi$	$-\tan\alpha$	$\mp\cot\alpha$	$\pm\tan\alpha$	$\mp\cot\alpha$	$-\tan\alpha$
$\cot\varphi$	$-\cot\alpha$	$\mp\tan\alpha$	$\pm\cot\alpha$	$\mp\tan\alpha$	$-\cot\alpha$

NOTES

SYMBOLS

S.1

A Area, Cross-sectional area (cm^2)	R Support reaction (kN), Strength (MPa),
D Diameter (cm), Force (kN)	Radius (cm)
E Modulus of elasticity (MPa)	S Settlement (cm)
For steel: $E = 2 \cdot 10^5$ MPa	S Elastic section modulus
E_s Modulus of deformation of soil (MPa)	about the neutral axis (cm^3)
F_c Centrifugal force (kN)	S_x " about the x - x axis (cm^3)
G Shear modulus of elasticity (MPa)	S_y " about the y - y axis (cm^3)
For steel: G = 77221MPa	S_z " about the z - z axis (cm^3)
H Horizontal support reaction (kN)	T^0 Temperature $(^0C, ^0F)$
I Moment of inertia of section	V Shear (kN), Volume (cm^3, m^3)
about the neutral axis (cm^4)	W Weight (kN)
I_x " about the x - x axis (cm^4)	Z Plastic section modulus (cm^3), Force (kN)
I_y " about the y - y axis (cm^4)	c Cohesion (Pa)
I_z " about the z - z axis (cm^4)	e Eccentricity (cm)
I_p Polar moment of inertia (cm^4)	g Gravitational acceleration $(g = 9.81 m/sec^2)$
K_0 Coefficient of earth pressure at rest	i Radius of gyration (cm)
K_a Coefficient of active earth pressure	k_w Winkler's coefficient of subgrade (kN/cm^3)
K_p Coefficient of passive earth pressure	n Porosity (%)
K_{aE} Coefficient of seismic active earth pressure	p Horizontal distributed load (kN/m)
L Span length (m)	w Vertical distributed load (kN/m)
M Mass (kg)	σ Direct stress (Pa)
M Bending moment	τ Shear stress (Pa)
about the neutral axis $(kN \cdot m)$	τ_s Shear strength (Pa)
M_x " about the x - x axis $(kN \cdot m)$	γ Unit volume weight (kN/m^3)
M_y " about the y - y axis $(kN \cdot m)$	μ Poisson's ratio
M_z " about the z - z axis $(kN \cdot m)$	α Coefficient of linear expansion (1/grad)
M_D Dynamic bending moment $(kN \cdot m)$	ρ Unit mass (kg)
N Axial force (kN)	Δ Deflection (cm)
P Applied load (kN)	ϕ Angle of internal friction
P_e Euler's force (kN)	$\tan \phi$ Coefficient of friction

NOTES